# Vibration Engineering - Analysis, Control, and Utilization

*Edited by Ming Yuan*

Published in London, United Kingdom

Vibration Engineering - Analysis, Control, and Utilization
http://dx.doi.org/10.5772/intechopen.1006779
Edited by Ming Yuan

Contributors
Abdelmoumen Anis Bousahla, Adriana Munteanu, Cătălin Gabriel Dumitraş, Dragoş-Florin Chitariu, Florin Chifan, George-Gabriel Chiriac, Hany A. Shawky, Matthew Clontz, Mihaiţă Horodincă, Ming Yuan, Ying He, Zahra Khoddami Maraghi



First published in London, United Kingdom, 2025 by IntechOpen
IntechOpen is the global imprint of INTECHOPEN LIMITED, registered in England and Wales, registration number: 11086078, 167-169 Great Portland Street, London, W1W 5PF, United Kingdom

For EU product safety concerns: IN TECH d.o.o., Prolaz Marije Krucifikse Kozulić 3, 51000 Rijeka, Croatia, info@intechopen.com or visit our website at intechopen.com.

British Library Cataloguing-in-Publication Data
A catalogue record for this book is available from the British Library

Vibration Engineering - Analysis, Control, and Utilization
Edited by Ming Yuan
p. cm.
Print ISBN 978-1-83634-806-1
Online ISBN 978-1-83634-805-4
eBook (PDF) ISBN 978-1-83634-807-8

# Meet the editor

Prof. Ming Yuan received his B. S., M. S., and Ph.D. degrees from Nanjing University of Aeronautics and Astronautics in Nanjing, China. From 2013 to 2014, he served as a visiting researcher at the Laboratoire de Mécanique des Structures et des Systèmes Couplés (LMSSC), Conservatoire National des Arts et Métiers (CNAM) in France. He is currently an associate professor at Nanjing University of Posts and Telecommunications, where he leads the Measurement and Control Technology Department. His research focuses on noise and vibration control, the development of intelligent acoustic devices, and micro-nano energy harvesting. He has published more than thirty peer-reviewed academic papers and has served as a reviewer for several prestigious academic journals, such as Nature Communications, Advanced Materials, and Applied Physics Letters.

# Contents

# Preface

Vibration engineering plays a pivotal role in comprehending and governing the behavior of dynamic systems across a wide array of applications. Spanning from the design of advanced materials and structures to the optimization of manufacturing processes and the advancement of control technologies, vibration analysis lies at the core of engineering innovation.

The book *Vibration Engineering – Analysis, Control, and Utilization* consolidates state-of-the-art research and methodologies from various domains within vibration engineering, offering an authoritative resource for academics and practitioners.

This book comprises a meticulous selection of six chapters, each addressing critical aspects of vibration analysis, control, and application. In the opening chapter, Ying He and Ming Yuan comprehensively examine the dynamic analysis of resonance-based linear piezoelectric actuators, a cornerstone topic in developing precision devices. Zahra Khoddami Maraghi delves into the intricate subject of vibration and instability in smart-composite sandwich structures, encompassing both flutter and divergence phenomena, which are indispensable to aerospace and automotive applications.

The contribution by Cătălin Gabriel Dumitraş and colleagues explores vibration in metal cutting processes, a fundamental aspect of manufacturing technology. Matthew Clontz's investigation into frequency-based substructuring elucidates the coupling and decoupling mechanisms of dynamic systems, providing invaluable insights for system design and optimization. Abdelmoumen Anis Bousahla presents a rigorous study on the free vibration analysis of functionally graded plates on viscoelastic foundations, a significant area in structural engineering and material science.

Lastly, Hany Metry's research on noise exposure assessment in occupational settings highlights the intersection of vibration engineering and human health, a progressively pertinent issue in contemporary work environments. Each chapter offers distinct perspectives, integrating theoretical analysis, experimental techniques, and practical applications.

We aspire that this book serves as an indispensable reference for professionals engaged in vibration engineering and as a catalyst for further research and development. The contributions contained herein exemplify the ongoing evolution of vibration science and its seamless integration with modern technologies.

We express our deepest gratitude to the authors for their remarkable contributions and to the readers for their unwavering interest in this dynamic and continuously advancing field.

**Ming Yuan**
College of Automation and Artificial Intelligence,
Nanjing University of Posts and Telecommunications,
Nanjing, China

Chapter 1

# Numerical Modeling and Dynamic Analysis of Resonance-Based Linear Piezoelectric Actuators

*Ying He and Ming Yuan*

## Abstract

This chapter presents a comprehensive analysis of resonance-based linear piezoelectric actuators, focusing on their numerical modeling, dynamic behavior, and driving mechanism. The chapter begins by introducing the fundamental principles and classification of these actuators, followed by an in-depth exploration of electromechanical coupling in the stator and the dynamic response of the mover. A key aspect of the analysis is the development of an integrated dynamic model that combines finite element analysis of the stator with analytical methods to model friction contact and mover motion. This model captures the complex interactions at the stator-mover interface, including the frictional forces that play a crucial role in actuator performance. To evaluate the system's performance, modal analysis, harmonic response analysis, and transient dynamic simulations are conducted. The steady-state behavior of the actuator, including output velocity and load transfer, is studied under varying operating conditions such as excitation voltage, phase difference, and pre-pressure. This comprehensive modeling approach provides valuable insights into the driving mechanism of Resonance-based linear piezoelectric actuators (RLPA), enabling accurate prediction of output performance. The findings contribute to guiding structural optimization, improving actuator efficiency, and informing control strategies for future RLPA applications.

**Keywords:** vibration utilization, piezoelectric actuator, electromechanical coupling, numerical modeling, friction contact

## 1. Introduction

Resonance-based linear piezoelectric actuators (RLPA) have attracted significant attention in recent years due to their high precision, compact size, and efficient actuation capabilities [1–3]. These actuators leverage the resonant vibrations of a stator, typically made of piezoelectric materials, to generate linear motion in a coupled mover system [4–6]. The development and optimization of RLPA require a comprehensive understanding of their dynamic characteristics and the interactions between the stator, mover, and frictional contact at the interface [7–12].

This chapter focuses on the numerical modeling and dynamic behavior of RLPA, providing a systematic approach to their analysis. It begins with an overview of the fundamental principles behind RLPA and their classification based on the resonance modes of the stator [13–15]. A key aspect of RLPA operation lies in the electromechanical coupling that governs the interaction between the stator and the driving mechanism. Understanding the dynamics of both the stator and the mover is crucial for accurate predictions of the actuator's performance.

A dynamic model of RLPA is developed by combining finite element analysis (FEA) for the stator with analytical models for frictional contact and the motion of the mover. This approach enables an in-depth examination of the actuator's modal characteristics, harmonic response, and transient dynamics. Through this modeling framework, the chapter addresses the essential dynamic coupling between the stator vibration, frictional driving force, and the motion of the mover. By studying the frictional driving mechanism at the interface between the stator and the mover, we gain valuable insights into the actuation process, which is crucial for both performance prediction and design optimization.

The goal of this chapter is to provide a foundation for further developments in RLPA technology, offering tools for the optimization of actuator designs and the development of control strategies aimed at practical applications in various fields. Through this analysis, we seek to enhance the efficiency and reliability of RLPA, paving the way for their broader adoption in high-precision motion control systems.

## 2. Classification and operating mechanism of RLPA

### 2.1 Classification based on stator resonance mode

Resonance-based linear piezoelectric actuators can achieve macroscopic linear output directly by utilizing the frictional contact between the stator and mover without the need for gears or other mechanical devices. This design significantly simplifies the actuator structure, enhancing the system's compactness and reliability. To avoid noise generation during operation, the working frequency is typically set in the ultrasonic range (above 20 kHz), which is why these actuators are also referred to as linear ultrasonic motors [16, 17].

Linear piezoelectric actuators exhibit flexible designs and diverse structures, and currently, there is no unified or systematic classification method. The primary distinction among these actuators lies in the vibration mode of the stator, as the vibration characteristics of the stator directly reflect the operating characteristics of the actuator. Therefore, RLPA can be classified into single-mode and composite-mode types based on the predominant vibration modes of the stator during operation.

- *Single-mode drive*: Single-mode drive refers to actuators in which the stator operates in a single specific resonance mode [18, 19]. Single-mode drives include longitudinal, bending, and torsional modes, each corresponding to different forms of vibration and driving characteristics.

- *Composite-mode drive*: Composite-mode drive refers to actuators where the stator operates simultaneously in two or more resonance modes [20, 21]. By combining multiple vibration modes, composite-mode drives achieve more complex driving mechanisms, resulting in improved overall performance, greater thrust, and increased movement versatility. Composite modes can include combinations such as longitudinal/bending, longitudinal/torsional, and bending/bending modes.

## 2.2 Driving mechanism based on elliptical trajectory

The operating principle of RLPA involves two distinct energy conversion processes. **Figure 1** presents a simplified schematic of the driving mechanism [17]. Initially, the positive piezoelectric effect of the piezoelectric ceramics excites the resonance mode of the stator, which has a specific structural design. This excitation causes the driving tip to produce microscopic vibrations in two perpendicular directions, with a phase difference between them. If the tangential displacement $u_x$ and the normal displacement $u_y$ of the driving tip are described by:

$$\begin{cases} u_x = U_x \sin(\omega t) \\ u_y = U_y \sin(\omega t + \theta) \end{cases} \tag{1}$$

where $U_x$ and $U_y$ represent the amplitude of the tangential and normal displacements, respectively, and $\theta$ is the phase difference between the displacement responses in the two perpendicular directions.

The motion trajectory of the driving tip at the contact interface with the mover can be derived by eliminating the time variable $t$ from the parametric equations Eq. (1). The resulting equation describes the elliptical trajectory of the driving tip:

$$\left(\frac{u_x}{U_x}\right)^2 + \left(\frac{u_y}{U_y}\right)^2 - 2\frac{u_x u_y}{U_x U_y}\cos(\theta) = \sin^2(\theta) \tag{2}$$

It can be seen that the elliptical trajectory of the driving tip is simultaneously influenced by the displacement amplitudes ($U_x$ and $U_y$) and the phase difference ($\theta$).

Secondly, the elastic force generated by the preload spring, combined with the normal movement of the driving tip, contributes to the contact pressure at the interface. Meanwhile, the tangential reciprocating motion produces work through frictional forces. This differential work, resulting from friction, enables the microscopic

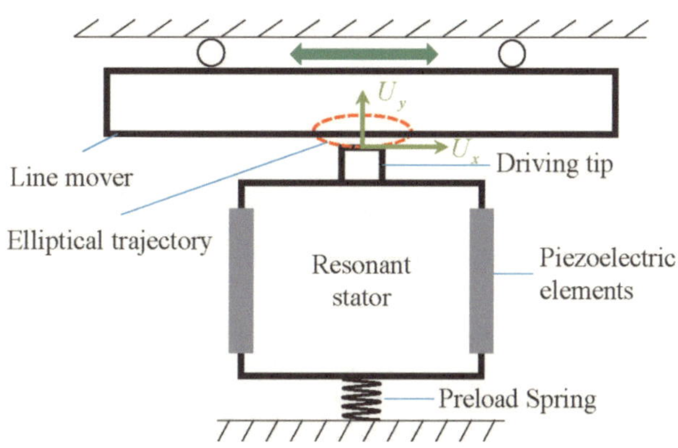


**Figure 1.**
*Illustration of driving process in RLPA.*

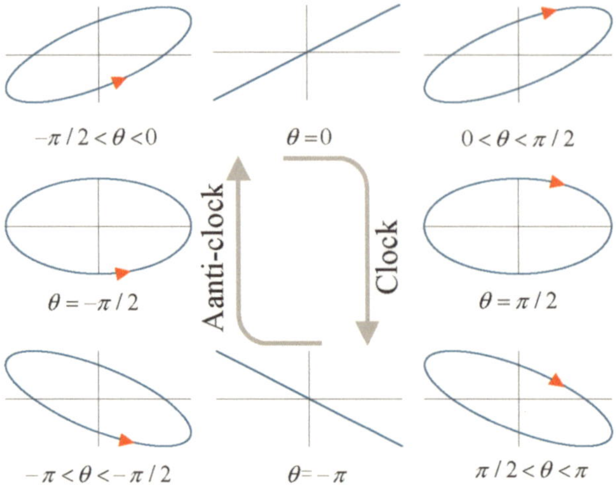


**Figure 2.**
*Elliptical motion trajectory of the driving foot for different phase differences θ.*

vibration of the stator's driving tip to be converted into the macroscopic continuous linear motion of the mover.

The friction force at the driving interface is determined by the amplitude of the normal vibration, and thus, the normal vibration amplitude largely dictates the output thrust of the actuator. On the other hand, the tangential vibration amplitude determines the amount of work done by the stator on the mover during a driving cycle, thereby affecting the actuator's output efficiency. The phase difference between the normal and tangential displacement responses affects both the shape and the direction of the elliptical trajectory. **Figure 2** illustrates the changes in the elliptical motion trajectory as described by Eq. (2) for different phase differences.

As the phase difference varies, significant changes occur in the inclination angle, rotation direction (clockwise or counterclockwise), and overall shape of the elliptical trajectory. These variations directly influence the movement direction and trajectory shape of the driving foot, thereby affecting the friction characteristics between the mover and the stator, ultimately impacting the output thrust and energy efficiency of the actuator. Therefore, designing the stator structure and tuning the electrical excitation parameters to optimize the phase difference between the normal and tangential displacements are crucial for achieving efficient driving and maximizing output performance.

## 3. Dynamic modeling and efficient solution of RLPA

### 3.1 Dynamic modeling of piezoelectric composite stator

The stator is a piezoelectric composite structure formed by integrating multiple piezoelectric elements into a specific elastic substrate. Due to the electromechanical coupling properties of piezoelectric materials and the geometric diversity of the elastic substrate, dynamic modeling of the stator is challenging to achieve analytically. This subsection presents an electromechanical coupled dynamic finite element modeling approach for the piezoelectric composite stator based on the energy method.

The focus is on capturing the essential electromechanical coupling and resonance characteristics required for its operation.

Based on Hamilton's principle [22, 23], the motion of an arbitrarily structured stator can be described by:

$$\delta\int_{t_1}^{t_2}[T-(U_M-W_M)+(U_E-W_E)]\,\mathrm{d}t=0, \tag{3}$$

Where $L = T - (U_M - W_M) + (U_E - W_E)$ is called the Lagrangian function. In this expression, $T$ represents the kinetic energy of the system, $U_M$ denotes the elastic strain energy, $W_M$ represents the work done by external mechanical forces, $U_E$ is the electric energy in the electric field, and $W_E$ denotes the work done by external electric forces.

For the piezoelectric region, each energy term [24] is defined as follows:

$$\begin{cases} T=\frac{1}{2}\iiint_V \rho\dot{\boldsymbol{u}}^T\dot{\boldsymbol{u}}dV, \\ U_M=\frac{1}{2}\iiint_V \boldsymbol{\sigma}^T\boldsymbol{\varepsilon}dV, \\ W_M=\iiint_V \boldsymbol{f}^T\boldsymbol{u}dV+\iint_\Sigma \boldsymbol{F}^T\boldsymbol{u}dS, \\ U_E=\frac{1}{2}\iiint_V \boldsymbol{E}^T\boldsymbol{D}dV, \\ W_E=\iiint_V \varphi q dV+\iint_\Sigma \varphi\rho_o dS, \end{cases} \tag{4}$$

If the electric-related terms are removed, the above expressions correspond to the elastic region of the material. In these equations, $V$ represents the integration volume while $\Sigma$ represents the boundary surface of volume $V$. The symbol $\rho$ denotes the material density, $\dot{\boldsymbol{u}}$ represents the velocity vector, which is the time derivative of the displacement vector $\boldsymbol{u}$. The stress tensor is represented by $\boldsymbol{\sigma}$, and the strain vector is represented by $\boldsymbol{\varepsilon}$. The body force density is denoted by $\boldsymbol{f}$, and the surface force density acting on the boundary surface $\Sigma$ is represented by $\boldsymbol{F}$. The electric field vector is $\boldsymbol{E}$, and the electric displacement vector is $\boldsymbol{D}$. The electric potential is represented by $\varphi$, the free body charge density is denoted by $q$, and $\rho_o$ represents the free surface charge density.

The energy components described above are calculated for the stator using the finite element method after discretizing the geometric structure of the stator. The piezoelectric and elastic regions are discretized into $e_1$ and $e_2$ elements. If the Lagrangian function within the i-th element is expressed as $L_i$. The Lagrange equation of the whole stator can be expressed as the sum of the individual elements, i.e.,

$$L=\sum_{i=1}^{e_1}L_i+\sum_{i=1}^{e_2}L_i \tag{5}$$

Substitute Eq. (5) into Eq. (3), the discrete governing equation for the dynamic system can then be expressed as:

$$\int_{t_1}^{t_2}\left(\sum_{i=1}^{e_1}\delta L_i+\sum_{i=1}^{e_2}\delta L_i\right)dt=0, \tag{6}$$

By substituting the constitutive and geometric equations of both the elastic substrate and the piezoelectric region into the above expression and performing interpolation over the subdivided elements, the Lagrangian equation $L_i$ for the i-th element is transformed into a representation involving nodal displacements and electric potentials for computational purposes. Specifically, the piezoelectric constitutive equations are expressed as:

$$\boldsymbol{\sigma} = \boldsymbol{c\varepsilon} - \boldsymbol{eE}, \quad \boldsymbol{D} = \boldsymbol{e}^{\mathrm{T}}\boldsymbol{\varepsilon} + \boldsymbol{gE}, \tag{7}$$

where $\boldsymbol{c}$, $\boldsymbol{e}$, and $\boldsymbol{g}$ are the elastic stiffness matrix, the piezoelectric stress matrix, and the permittivity matrix of the piezoelectric ceramic material, respectively. These matrices characterize the material properties and have dimensions of $6 \times 6, 6 \times 3,$ and $3 \times 3$, respectively. The piezoelectric geometric equations are expressed as:

$$\boldsymbol{\varepsilon} = \boldsymbol{Lu}, \quad \boldsymbol{E} = -\nabla\phi, \tag{8}$$

where $\boldsymbol{L}$ and $\nabla$ represent the differential operator and the gradient operator, respectively. For purely elastic materials, the corresponding equations can be obtained by removing the electric-related terms.

The discretization process uses shape functions to interpolate the displacements and electric potentials within each element, as follows:

$$\boldsymbol{u}(x,y,z) = \boldsymbol{N}\hat{\boldsymbol{u}}, \quad \phi(x,y,z) = \boldsymbol{N}_\phi\hat{\phi}, \tag{9}$$

where $\hat{\boldsymbol{u}}$ and $\hat{\phi}$ are the nodal displacement and potential vectors, respectively, and $\boldsymbol{N}$ and $\boldsymbol{N}_\phi$ are the shape functions. This formulation enables the governing equations to be assembled for the entire system.

Based on Eqs. (7), (8), and (9), the energy terms in Eq. (4) are expressed in terms of nodal displacements and potentials. Subsequently, the element energy term is substituted into Eq. (6) to perform the variation calculation. As a result, the matrix form of the finite element dynamic equation for the stator can be derived as follows:

$$\begin{bmatrix} \mathbf{M}_{uu}^1 & 0 & 0 \\ 0 & \mathbf{M}_{uu}^2 & 0 \\ 0 & 0 & 0 \end{bmatrix} \begin{bmatrix} \ddot{\mathbf{u}}^1 \\ \ddot{\mathbf{u}}^2 \\ \ddot{\mathbf{\Phi}} \end{bmatrix} + \begin{bmatrix} \mathbf{K}_{uu}^1 & 0 & 0 \\ 0 & \mathbf{K}_{uu}^2 & \mathbf{K}_{u\phi}^2 \\ 0 & \mathbf{K}_{\phi u}^2 & \mathbf{K}_{\phi\phi}^2 \end{bmatrix} \begin{bmatrix} \mathbf{u}^1 \\ \mathbf{u}^2 \\ \mathbf{\Phi} \end{bmatrix} = \begin{bmatrix} \mathbf{F}^1 \\ \mathbf{F}^2 \\ \mathbf{Q} \end{bmatrix} \tag{10}$$

Here, superscripts 1 and 2 represent the elastic region and the piezoelectric region, respectively. $\mathbf{u}$ represents the nodal displacement vector, and $\mathbf{\Phi}$ represents the electric potential vector. $\mathbf{M}_{uu}$ represents the mass matrix, and $\mathbf{K}_{uu}$ denotes the mechanical stiffness matrix. The coupling matrices $\mathbf{K}_{u\phi}$ and $\mathbf{K}_{\phi u}$ describe the electromechanical coupling of piezoelectric material. $\mathbf{K}_{\phi\phi}$ represents the dielectric stiffness matrix for the piezoelectric region. $\mathbf{F}$ and $\mathbf{Q}$ represent the equivalent nodal load vectors, derived from the assembly of external forces and charges applied to the system.

Using the direct displacement and electric potential constraint method [25], fixed constraints are applied to specific nodes of the stator along with electrical signal excitation. By directly adjusting the values of the constrained displacement and electric potential degrees of freedom in Eq. (10) and reorganizing the matrix equation, the finite element dynamic equation of the stator incorporating the constraints can be obtained:

$$\mathbf{M}\ddot{\mathbf{a}} + \mathbf{K}\mathbf{a} = \mathbf{F}_{\text{pre}} + \mathbf{F}_{\text{e}}, \tag{11}$$

where $\mathbf{a} = [\mathbf{u}^1 \ \mathbf{u}^2 \ \mathbf{\Phi}]^T$ presents the unknown node displacement and potential. $\mathbf{F}_{\text{pre}}$ and $\mathbf{F}_{\text{e}}$ represent the equivalent nodal force vectors, which are calculated from the pre-pressure applied during finite element model assembly and the electrical potential, respectively.

To account for damping during the vibration process of the stator, Rayleigh damping $\mathbf{C} = \alpha\mathbf{M} + \beta\mathbf{K}$ is adopted. Finally, considering damping, the governing equation for the finite element model of the stator can be written as:

$$\mathbf{M}\ddot{\mathbf{a}} + \mathbf{C}\dot{\mathbf{a}} + \mathbf{K}\mathbf{a} = \mathbf{F}_{\text{pre}} + \mathbf{F}_{\text{e}}. \tag{12}$$

To optimize the structure of the stator, it is essential to first determine its operational mode and frequency. The $i$-th natural angular frequency $\omega_i$ of the stator and its corresponding mode shape $\varphi_i$ is obtained by solving the eigenvalue and eigenvector problem of the stator's dynamic equation Eq. (12). Specifically, this involves solving the following equation:

$$(\mathbf{K} - \omega_i^2\mathbf{M})\varphi_i = 0, \tag{13}$$

By solving this eigenvalue problem, the working mode order $i$ of the stator can be selected, and the corresponding natural frequency $f = \frac{\omega_i}{2\pi}$ can be used to estimate the operating frequency.

For the piezoelectric element of the stator, under the excitation of a harmonic voltage, it is often necessary to analyze the displacement responses of the driving tip in two perpendicular directions of the piezoelectric body in order to predict the shape and direction of the elliptical driving trajectory. This can be solved through harmonic analysis, with the governing equation represented as:

$$(\mathbf{K} - \omega^2\mathbf{M} + j\omega\mathbf{C})(\mathbf{a}_1 + j\mathbf{a}_2) = \mathbf{F}_{e1} + j\mathbf{F}_{e2}, \tag{14}$$

where $j = \sqrt{-1}$ represents the imaginary unit, $\omega$ is the angular frequency of the excitation voltage, is the real part of the equivalent load vector $\mathbf{F}_{\text{e}}(t)$ due to the excitation voltage, and $\mathbf{F}_{e2}$ is the imaginary part. By solving this complex linear system, the complex displacement $u = u_1 + ju_2$ at any degree of freedom of the driving tip can be obtained, which can then be represented in terms of the magnitude $|u|$ and the phase angle $\theta$ as follows:

$$|u| = \sqrt{u_1^2 + u_2^2}, \quad \theta = \arctan\left(\frac{u_2}{u_1}\right), \tag{15}$$

By separately calculating the response magnitudes in the normal and tangential directions as well as their phase differences, the elliptical motion trajectory of the driving tip can be determined.

### 3.2 Microscopic surface friction contact model

In the BLPA, the micro-vibration of the stator is converted into the macroscopic linear motion of the mover through frictional contact between the stator and the mover. The dynamic transmission model at the contact interface between the stator and the mover determines the final output characteristics of the actuator. Due to the high-frequency, small-amplitude vibration of the stator's driving interface and the variability of vibration amplitude, it is crucial to establish a surface friction contact model that takes into account the microscopic morphology of the interface in order to accurately describe the complex friction contact mechanism between the stator and the mover.

According to friction contact theory, the load transmission at the contact interface between the stator and mover involves both normal and tangential components. To establish a mathematical model for load transmission in these directions while considering the microscopic morphology of the contact interface, the two contacting surfaces are initially simplified as a smooth plane and an equivalent rough surface, as illustrated in **Figure 3**. The equivalent rough surface is approximated as a series of asperities with a uniform curvature radius $R$, with asperity height ($z$) following a Gaussian random distribution. For a rough surface with roughness $l$, the probability density function of the height can be expressed as:

$$\psi(z) = \left(\frac{1}{2\pi l^2}\right)^{1/2} e^{-\frac{z^2}{2l^2}}, \tag{16}$$

Based on the Greenwood-Williamson (G-W) normal contact model [26], each asperity on the equivalent rough surface is assumed to undergo elastic deformation during contact. The average normal contact pressure $p$ of the asperities per unit area, represented as a function of indentation depth, is given by:

$$p = \frac{4}{3}\eta E\left(\frac{R}{2\pi l^2}\right)\int_{h_{\max}}^{h_{\max}+\delta} e^{-\frac{x^2}{2l^2}}(x - h_{\max} + \delta)^{\frac{3}{2}}dx, \tag{17}$$

where $\eta$ is the asperity density. $E$ is the elasticity modulus. $h_{\max}$ is the maximum asperity height, which follows a Gaussian distribution such that 99.9% of the heights are within [3 $l$, 3 $l$]. Therefore, the maximum height is given by $h_{\max} = 3l.\delta$ is the indentation depth, which, in cases where the normal displacement of the mover's

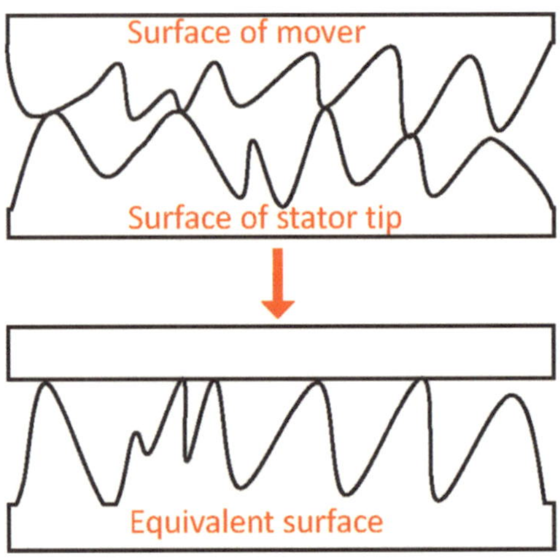


**Figure 3.**
*Normal contact model based on microscopic surface morphology.*

contact interface is neglected, can be represented as the normal displacement $u_s$ of the stator's driving surface during contact, given by:

$$\delta = \begin{cases} u_s & u_s > 0 \\ 0 & u_s \leq 0 \end{cases} \tag{18}$$

The normal nodal force vector $\mathbf{F}_n^e$ for each contact element on the driving tip can be calculated by:

$$\mathbf{F}_n^e = \int_S \mathbf{N}^T \mathbf{p} dS, \tag{19}$$

where $\mathbf{N}$ is the shape function matrix of the surface element, and $\mathbf{p}$ represents the contact pressure on the engaged contact element. The overall equivalent nodal force $\mathbf{F}_n$ on the driving interface can be assembled from the equivalent nodal forces of all contact elements. This force is then placed on the right-hand side of Eq. (12), adding the normal contact forces acting on the stator driving interface into the stator's dynamic system.

To characterize static friction, viscous friction, and hysteresis during the stator-mover interaction, the LuGre dynamic friction model [27, 28] is adopted. The tangential friction force for any arbitrary node $i$ on the contact interface is expressed as:

$$F_{ti} = E_0 w + E_1 \dot{w}, \tag{20}$$

Here, the state variable $w$ represents the average tangential deformation, $E_0$ represents the equivalent shear modulus of the contact interface, $E_1$ donate tangential damping coefficient. The average deformation state is computed dynamically as:

$$\dot{w} = v_r - \frac{E_0}{g(v_r)} w, \tag{21}$$

where $g(v_r)$ is a function of the relative sliding velocity at the interface, representing dynamic friction behavior. The static friction surface force $F_{t,c}$ and sliding friction surface force $F_{t,s}$ are given by:

$$F_{ti,c} = \mu_c F_{ni}, \quad F_{ti,s} = \mu_s F_{ni}, \tag{22}$$

where $\mu_c$ and $\mu_s$ are the static and sliding friction coefficients, respectively, and $F_n i$ represents the normal pressure acting at the contact node.

The function $g(v_r)$ is defined as:

$$g(v_r) = F_{ti,c} + (F_{ti,s} - F_{ti,c}) e^{-(v_r/v_s)^\alpha}, \tag{23}$$

where $v_r$ is the relative sliding velocity between the contact surfaces, $v_s = 1 \times 10^{-3}$ is a reference velocity, and $\alpha = 2$ is an adjustable parameter used to describe the influence of velocity on friction force.

By solving the deformation state equations for the contact interface at any arbitrary node, the corresponding tangential friction force can be calculated. Then, the tangential friction nodal force is then assembled into the overall equivalent nodal force matrix $\mathbf{F}_t$ for tangential friction. This force is then placed on the right-hand side of Eq. (12), adding the tangential friction forces acting on the stator driving interface into the stator's dynamic system.

### 3.3 Mover dynamic modeling

In the BLPA, the mover is considered to move linearly along the guiding direction under the frictional driving force exerted by the stator without experiencing deformation or vibration within the operating frequency range of the stator. Therefore, the mover, under the influence of the frictional driving force from the stator and the frictional resistance from the guide rail, can be modeled as a single-degree-of-freedom (SDOF) system, as shown in **Figure 4**. Here, $F_{\text{load}}$ represents the external load applied to the actuator, $F_n$ and $F_t$ represent the total normal and tangential forces at the mover-stator contact interface, respectively, and $F_f$ represents the frictional force between the mover and the guide rail. $m$ is the mass of the mover and $g = 9.8\, m/s^2$ is the gravitational acceleration.

Based on the force equilibrium relationship and considering the energy dissipation through the viscous damping coefficient $c$ during the frictional driving process, the dynamic model of the mover can be expressed as:

$$m\ddot{u}_m + c\dot{u}_m = F_t + F_f - F_{\text{load}}. \tag{24}$$

The frictional force between the mover and the guide rail is given by:

$$F_f = \text{sign}(-u_m)\mu_m(mg + F_n), \tag{25}$$

where $\mu_m$ is the friction coefficient between the guide rail and the mover, and $\text{sign}(-u_m)$ indicates the direction of the frictional force.

The frictional contact forces exerted by the stator on the mover can be obtained by summing the equivalent nodal forces over the driving interface of the stator:

$$F_t = -\sum \mathbf{F}_t(i), \quad F_n = -\sum \mathbf{F}_n(i), \tag{26}$$

### 3.4 Efficient solution of the system dynamic model

To conduct a comprehensive dynamic analysis of the BLPA, it is essential to consider the coupled interactions among the stator, mover, and frictional contact. Accordingly, the frictional contact model derived in the previous sections is

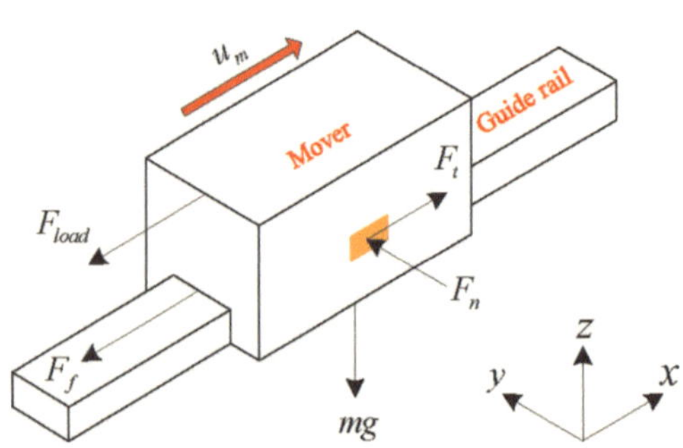


**Figure 4.**
*Schematic representation of mover dynamics and applied forces.*

incorporated into the finite element model of the stator and the single-degree-of-freedom model of the mover. Specifically, the equivalent nodal forces of the normal and tangential frictional contact at the driving interface, $\mathbf{F}_{\mathrm{n}}$ and $\mathbf{F}_{\mathrm{t}}$, obtained in Section 3.2 are added to the stator finite element model Eq. (12), resulting in a discrete dynamic model of the stator that accounts for frictional contact effects. By coupling this model with the single-degree-of-freedom mover system model Eq. (24) and the tangential deformation state equation Eq. (21) at the interface, the dynamic model of the coupled system can be established as follows:

$$\begin{cases} \mathbf{M}\mathbf{a} + \mathbf{C}\dot{\mathbf{a}} + \mathbf{K}\mathbf{a} = \mathbf{F}_{\mathrm{pre}} + \mathbf{F}_{\mathrm{e}} + \mathbf{F}_{\mathrm{t}} + \mathbf{F}_{\mathrm{n}} \\ m\ddot{u} + c\dot{u} = F_t + F_f - F_{\mathrm{load}} \\ \dot{\mathbf{w}} = \mathbf{v}_{\mathrm{r}} - \mathbf{h}\mathbf{w} \end{cases} \tag{27}$$

The third equation represents the assembly of the system of Eq. (21) used to solve the tangential deformation at the friction interface for each node. In this context, $\mathbf{h}$ is a diagonal matrix with diagonal elements $h_i = E_0 \frac{|v_{ri}|}{g(v_{ri})}$, where $i$ is the node index at the contact interface.

The system's dynamic response can theoretically be obtained by solving the above equations using numerical integration. However, due to the high-frequency characteristics of the actuator and the strong nonlinearity of the frictional contact, a direct solution of these equations would require substantial computational resources. Given that the BLPA operates at specific resonance modes, the unknown state vector $\mathbf{a}$ is represented in terms of modal coordinates $\mathbf{q}$:

$$\mathbf{a} = \mathbf{\Lambda}\mathbf{q} \tag{28}$$

where $\mathbf{\Lambda}$ is the modal shape matrix, which can be written as:

$$\mathbf{\Lambda} = \begin{bmatrix} \varphi_1 & \varphi_2 & \cdots \end{bmatrix} \tag{29}$$

Using the orthogonality of modal coordinates [29], the system's dynamics equation Eq. (27) is transformed into modal coordinates, yielding the following system:

$$\begin{cases} \mathbf{q} + (\alpha\mathbf{I} + \beta\lambda)\mathbf{q} + \lambda\mathbf{q} = \mathbf{\Lambda}^T\left(\mathbf{F}_{\mathrm{pre}} + \mathbf{F}_{\mathrm{e}} + \mathbf{F}_{\mathrm{t}} + \mathbf{F}_{\mathrm{n}}\right) \\ m\ddot{u} + c\dot{u} = F_t + F_f - F_{\mathrm{load}} \\ \dot{\mathbf{w}} = \mathbf{v}_{\mathrm{r}} - \mathbf{h}\mathbf{w} \end{cases} \tag{30}$$

where $\mathbf{I}$ is the identity matrix and $\lambda = \underset{1 \le i \le N}{\operatorname{diag}} [\omega_i]$, with $N$ being the number of modes considered. By focusing on the actuator's resonant modes, the order of the system can be significantly reduced.

For the dynamic equations Eq. (30) in this modal space, even when using the high-precision fourth-order Runge–Kutta method for time-domain integration, rapid simulation of the full-system dynamics can be performed on standard computational hardware.

## 4. Dynamic simulation of a V-shaped RLPA

Taking a V-shaped RLPA [30] as an example, the dynamic characteristics of the stator and the output performance of the actuator are analyzed based on the proposed

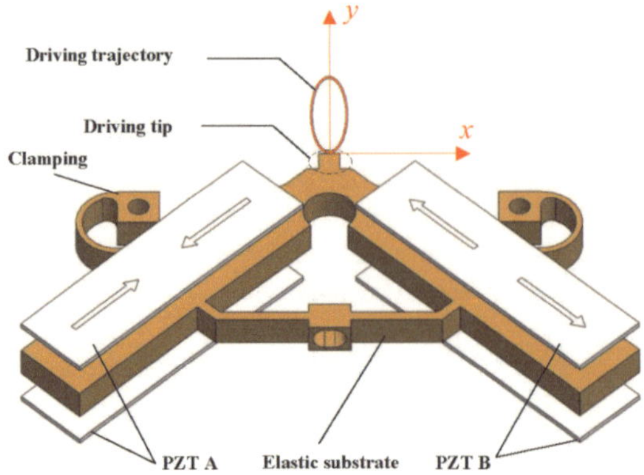


**Figure 5.**
*Schematic diagram of the V-shaped RLPA stator structure.*

model. **Figure 5** shows the schematic of the V-shaped stator structure. Four rectangular PZT-8 piezoelectric elements are bonded to the V-shaped beryllium copper elastic substrate using epoxy resin. Two sinusoidal electric fields with a phase difference are applied to the piezoelectric element arrays on both sides. In the case of the two piezoelectric elements on the stator left, one side is grounded, and a sinusoidal voltage signal $\varphi_1$ with amplitude $V$, frequency $f$, and zero phase is applied to the other side:

$$\varphi_1 = V\sin(2\pi ft) \tag{31}$$

On the stator right, the configuration is identical, with one side grounded. A sinusoidal voltage signal $\varphi_1$ with the same amplitude and frequency as on the left side is applied, but with a phase difference $\hat{\theta}$:

$$\varphi_2 = V\sin\left(2\pi ft + \hat{\theta}\right) \tag{32}$$

The deformation of the piezoelectric elements caused by the applied electric field will excite specific resonance modes of the stator. The driving tip achieves an elliptical motion trajectory, which then drives the mover to produce macroscopic linear motion.

To validate the accuracy of the dynamic model in predicting the actuator's output performance, a comprehensive testing platform, shown in **Figure 6**, was constructed. This platform integrates several essential devices to simulate the operational environment and evaluate the system's dynamic response. The signal generator (AFG 3022B, Tektronix, Inc., USA) produces two-phase excitation signals, which serve as the input voltages applied to the piezoelectric elements. These signals, with adjustable amplitudes and frequencies, are amplified using a power amplifier (HFVA-42, Nanjing Funeng Technology Corp., China) to ensure that the piezoelectric elements receive sufficient driving voltage. The amplified signals directly drive the stator, inducing motion in the mover.

The output performance of the system is measured using a high-precision laser displacement sensor (LK-H150, Keyence Inc., Japan). This sensor monitors the linear motion of the mover in real time and sends the data to a PC terminal equipped with custom data processing software. The test platform also includes a load device attached to the mover, which applies varying external forces to simulate real-world operating conditions.

The testing process involves operating the system under various conditions to validate the dynamic model. Initially, the actuator is tested without an external load to evaluate its basic motion characteristics. Subsequently, different load forces are applied to the mover, and its response is recorded to assess the model's ability to predict the system's behavior under practical conditions. Additionally, frequency response tests are conducted by varying the excitation frequency to identify the system's resonant frequencies and compare them with theoretical predictions.

These experiments aim to ensure the accuracy of the dynamic model in predicting the mover's output speed and the overall system response. By comparing the test results with the model's predictions, the validity of the model is evaluated in terms of its ability to capture key dynamic characteristics, including the relationship between input and output and the effects of varying load forces. The parameters for the dynamic friction contact simulation at the stator-mover contact interface used in the system's dynamic model are shown in **Table 1**.

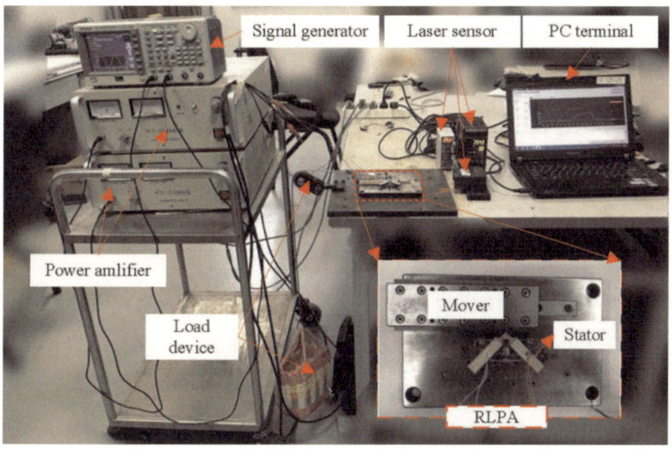


**Figure 6.**
*Experimental platform for output performance testing of the actuator.*

| Parameter | Description | Value | Unit |
|---|---|---|---|
| $R$ | Curvature radius | $1.6 \times 10^{-6}$ | mm |
| $l$ | Roughness | $6 \times 10^{-6}$ | μm |
| $\eta$ | Asperity density | $1.1 \times 10^{8}$ | $m^{-2}$ |
| $E$ | Elasticity modulus | $1.63 \times 10^{11}$ | Pa |
| $E_0$ | Shear modulus | $1.47 \times 10^{6}$ | Pa |
| $E_1$ | Tangential damping coefficient | $2.42 \times 10^{3}$ | $N \cdot s/m$ |
| $\mu_c$ | Static friction coefficients | 0.5 | — |
| $\mu_s$ | Dynamic friction coefficients | 0.35 | — |

**Table 1.**
*Parameters used for dynamic friction contact simulation.*

### 4.1 Analysis of the stator's dynamic characteristics

The stator, as the core component of the actuator, plays a critical role in the structural design and optimization of the RLPA. Analyzing the dynamic characteristics of the stator is an essential process in this context. By solving the characteristic equation of the stator's dynamic model, its resonant modes and frequencies can be quickly determined. Additionally, harmonic response analysis allows for the quantitative estimation of the displacement response amplitude at the driving end.

**Figure 7** shows the finite element mesh used to discretize the V-shaped stator for dynamic modeling. By solving the eigenvalue equation of the stator's dynamic model (Eq. 13), the two operating modes and their corresponding frequencies suitable for driving are identified. Under the action of the electric field excitation, the piezoelectric elements induce the stator to vibrate in both symmetric and anti-symmetric modes (mode A1 and mode A2). These vibrations cause the driving end to produce displacements with phase differences in both normal and tangential directions.

A voltage of 100 V with a 90° phase difference is applied to the piezoelectric elements on both sides. By solving the finite element dynamic model of the stator in the frequency domain (Eq. 14), harmonic response analysis is conducted to simulate the steady-state displacement amplitudes and phase differences of the normal and tangential displacements at the center of the driving tip. The frequency range is from 45 to 60 kHz, and the corresponding changes in the response curves are shown in **Figure 8**. From the displacement amplitudes, the operating frequency range of the actuator can be estimated. As for the phase differences, the normal and tangential displacement responses exhibit phase differences, which satisfy the condition for the formation of the elliptical driving trajectory.

### 4.2 Force transfer of the driving interface

In the RLPA, the frictional contact load between the stator and the mover directly determines the output of the mover. This section presents simulations and experiments on the transient dynamics of the V-shaped stator structure and analyzes the load transfer behavior of the frictional contact forces at the driving interface during the steady-state phase to help understand the driving mechanism of the actuator.

In the simulation, the time-domain response of the mover and the modal coordinate time-domain response of the stator is obtained by solving the dynamic equations

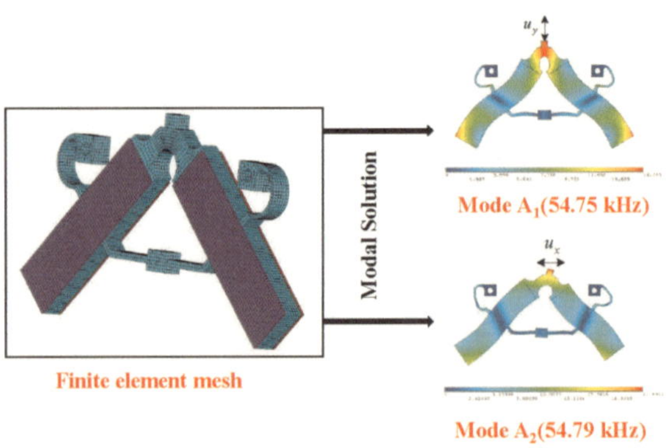


**Figure 7.**
*Finite element mesh and operating modes of the V-shaped stator dynamic model.*

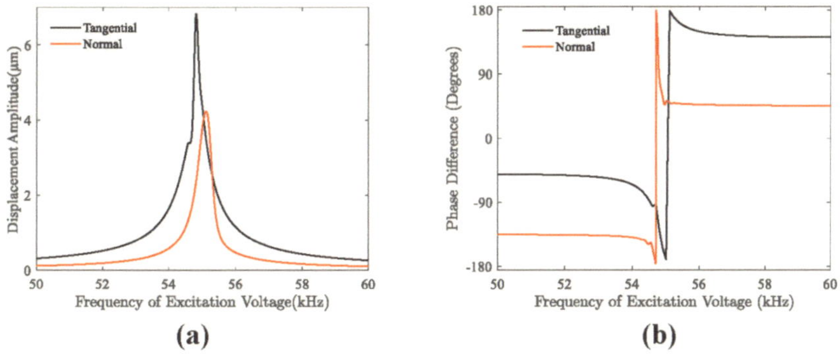


**Figure 8.**
*Steady-state response characteristics of driving tip: (a) Amplitude-frequency curve, (b) phase-frequency curve.*

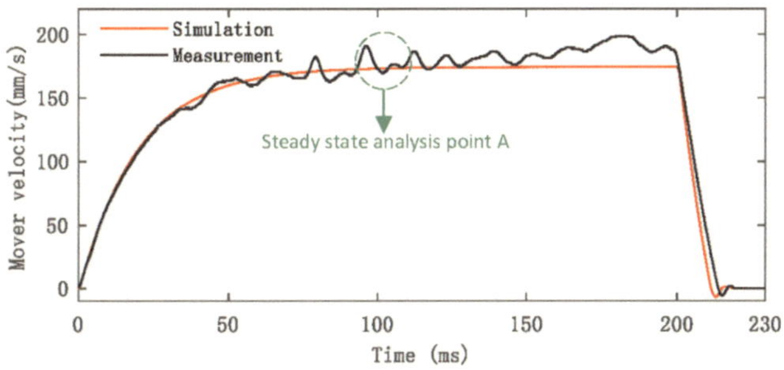


**Figure 9.**
*Transient velocity response of mover.*

Eq. (30) of the coupled system in the modal coordinates of the stator. Using the modal coordinate transformation in Eq. (28), the time-domain response of any finite element node on the stator can be calculated. By combining the transient responses of the mover and the stator at the driving interface with the frictional contact model described in Section 3.2, the normal and tangential load transfer at the stator-mover contact interface can be reconstructed.

When a sinusoidal voltage excitation with a frequency of 55 kHz, an amplitude of 100 V, and a phase difference of 90° is applied to the piezoelectric units on both sides of the V-shaped RLPA, along with a preloading condition of 10 N, **Figure 9** shows the simulation and experimental results of the transient response after applying voltage excitation to the piezoelectric unit for 200 ms, followed by power cutoff. The measured velocity shows some fluctuations, which are mainly caused by the positional differences in the microscopic morphology of the driving interface. Overall, the dynamic modeling method proposed in this paper shows good agreement with the experimental results in simulating the transient velocity.

After selecting the steady-state velocity point A of the actuator, the transient variations in normal pressure and tangential frictional force at the driving interface during one velocity cycle are calculated, as shown in **Figure 10(a)**. It can be observed that the output velocity of the actuator exhibits periodic fluctuations within a certain range, which is caused by the periodic changes in the interface frictional force. By combining the elliptical driving trajectory at the center of the driving tip, as shown in **Figure 10(b)**, a driving cycle can be divided into four stages: $t_1$-$t_2$, $t_2$-$t_3$, $t_3$-$t_4$, and $t_4$-$t_1$.

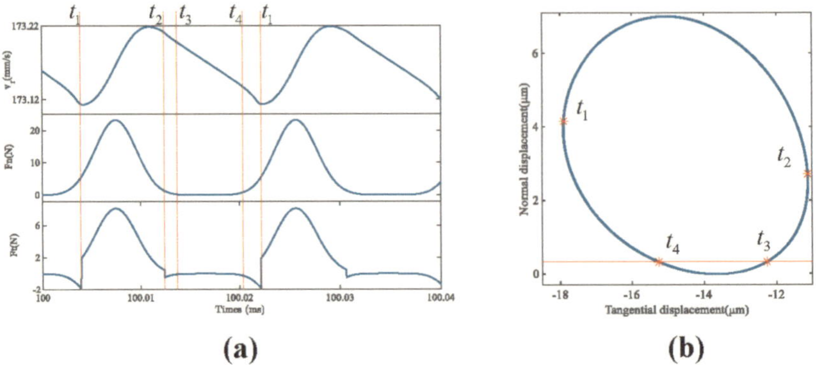

**Figure 10.**
*Steady-state driving process: (a) Load transfer at the driving interface, (b) corresponding elliptical driving trajectory.*

In the $t_1$-$t_2$ stage, the driving tip moves forward, doing positive work on the mover and causing the velocity to rise. As the tangential velocity of the driving tip drops below the mover's velocity, the friction force reverses and does negative work on the mover, causing the velocity to decrease. During the $t_2$-$t_3$ stage, the driving tip moves in reverse. In the subsequent $t_3$-$t_4$ stage, the contact interface between the stator and mover either separates or maintains contact with only a few asperities, resulting in the disappearance of frictional force. Consequently, the driving tip ceases to perform work on the mover. However, due to the effects of damping and guide rail friction, the mover's velocity continues to decrease. In the $t_2$-$t_3$ and $t_4$-$t_5$ stages, the driving tip does negative work on the mover, causing the velocity to decrease further.

In summary, the mover's velocity reaches a balanced range of fluctuations, which is due to the balance between the positive and negative work done by the driving tip during a single driving cycle and the damping force and guide rail friction that results from the velocity increase.

### 4.3 Output performance simulation

Based on the previously established system model and solution method, this section simulates and predicts the steady-state output characteristics of the V-shaped RLPA under different driving conditions. The focus is on analyzing the effects of various driving parameters, such as excitation voltage frequency, phase difference, pre-pressure, and output load, on the mover's output speed. Under different driving parameters, the equivalent nodal force vectors corresponding to the preload force $\mathbf{F}_{pre}$ on the stator and the electrical excitation $\mathbf{F}_e$ on the stator are calculated applied to the stator model of the coupled system dynamic equations Eq. (30). Additionally, the external load $F_{load}$ on the mover is applied to the mover's model. The coupled equations are then numerically solved to simulate the mover's output speed under various driving parameters. The excitation voltage is maintained at a constant amplitude of 100 V, while the other driving parameters are preset, as shown in **Table 2**, unless otherwise specified.

**Figure 11** shows the experimental and simulation results of the output speed variation with excitation voltage frequency in the range of 50–60 kHz. The speed variation follows a similar pattern to the frequency response curve of the driving tip, with a peak-like change. Notably, the frequency bandwidth of the output

| Parameter | Frequency (kHz) | Phase (degree) | Pre-pressures (N) | Output load (N) |
|---|---|---|---|---|
| Value | 55 | 90 | 10 | 0 |

**Table 2.**
*Preset driving parameters for the performance simulation of V-shaped RLPA.*

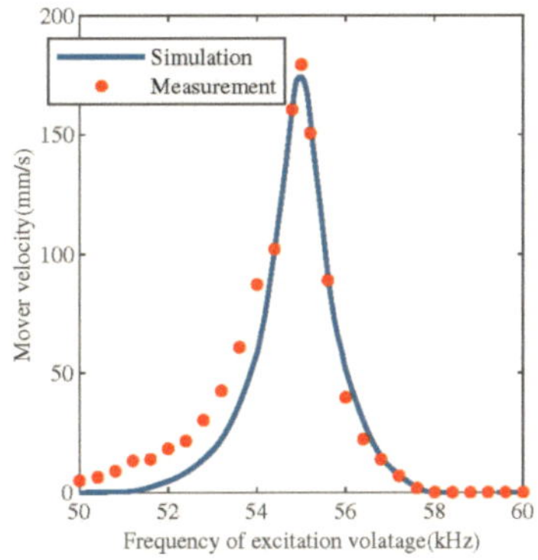


**Figure 11.**
*Output speed variation with frequency of excitation voltage.*

speed is smaller than the amplitude range of the driving tip (**Figure 8**), indicating that although the driving tip generates elliptical trajectories, it may not always drive the mover. The corresponding frequency is referred to as the dead zone frequency.

**Figure 12** illustrates the simulation and experimental results of the output speed variation with respect to the phase difference of the excitation voltage applied to both sides of the stator, with the phase difference range set from 0 to 360 degrees. The speed variation exhibits a quasi-sinusoidal pattern with respect to phase difference. The output speed is positive when the phase difference is between 0° and 170° and negative between 190° and 360°. A "dead zone" is observed between 170° and 190°, where the motor exhibits no output. This phase range is referred to as the dead zone phase. The speed peaks occur at phase differences of 60° and 300°, which are key points to consider when controlling the actuator by excitation voltage phase difference.

**Figure 13** shows the simulation and experimental results of the output speed variation with pre-pressure, with the pre-pressure range set from 0 to 45 N. As the pre-pressure increases, the speed initially increases linearly, then reaches two peaks before gradually decreasing. This is due to the increase in friction, which enhances the work done on the mover. However, the increase in pre-pressure also reduces the normal displacement, which decreases the work done during the elliptical driving cycle and leads to a reduction in speed. When the pre-pressure reaches a certain value, static friction occurs at the driving interface, increasing the friction coefficient and forming the second peak of speed.

**Figure 14** shows the simulation and experimental results of the output speed variation with external load, with the load range set from 0 to 4 N. As the external load increases, the speed decreases linearly to zero, and the motor is unable to drive the load, resulting in negative speed growth due to load drag.

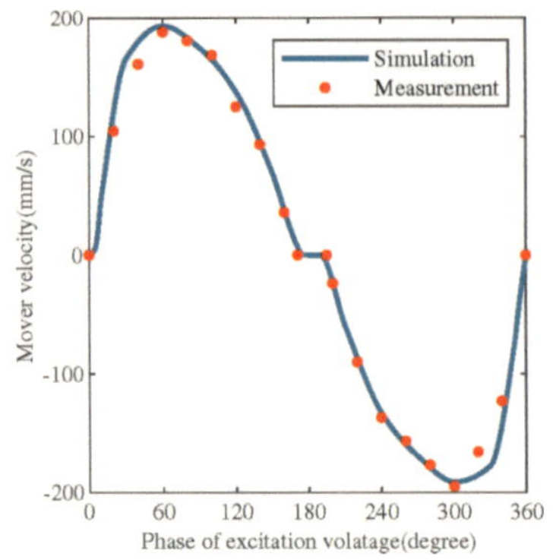


**Figure 12.**
*Output speed variation with phase of excitation voltage.*

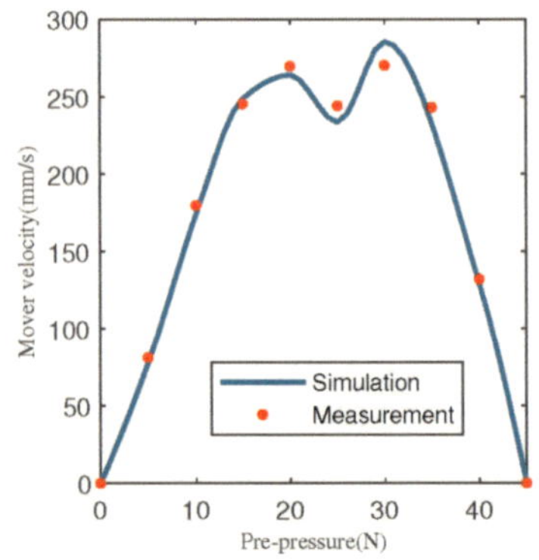


**Figure 13.**
*Output speed variation with pre-pressure of excitation voltage.*

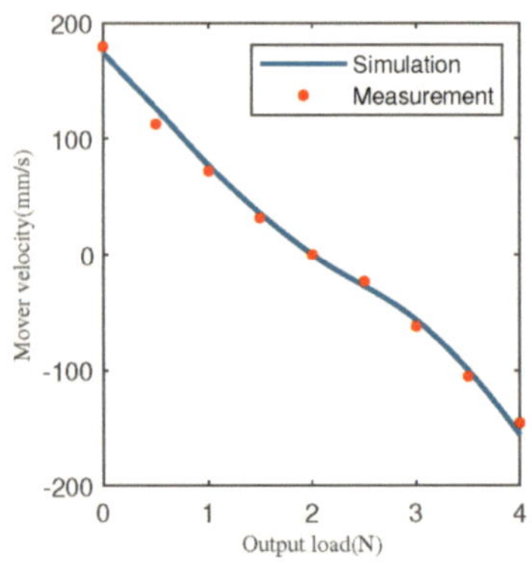


**Figure 14.**
*Output speed variation with output load of excitation voltage.*

## 5. Conclusions

This chapter develops a comprehensive dynamic model for resonance-based linear piezoelectric actuators, integrating finite element analysis for the stator, analytical frictional contact models, and mover dynamics. Using a V-shaped RLPA as a case study, the model effectively predicts both steady-state and transient behavior under varying operating conditions. The key contributions of this work include:

- Development of an integrated modeling framework that captures the dynamic coupling between stator vibration, frictional driving forces, and mover motion.

- Insights into the periodic frictional driving mechanism of RLPA, providing a deeper understanding of the interaction dynamics at the stator-mover interface.

- Validation of the model through experimental results, demonstrating its accuracy in predicting output velocity and load transfer characteristics.

By addressing the critical aspects of RLPA dynamics, this study lays the groundwork for future advancements in actuator technology. The findings provide a foundation for future studies aimed at optimizing RLPA designs to enhance efficiency and reliability, indicating their potential for applications in precision motion control systems.

Despite its contributions, the proposed model has certain limitations. For instance, the mover is assumed to behave as a rigid body, which may oversimplify the interaction between the stator and mover. Additionally, the influence of environmental factors, such as temperature and humidity, on frictional behavior was not considered. Furthermore, the dynamic friction model assumes uniform material properties, which may not fully capture variations in real-world applications. To address these limitations, future studies could explore:

- Refining the frictional contact model to account for additional factors influencing dynamic variations in contact forces, such as the anisotropy and non-uniformity of surface roughness.

- Investigating the effects of environmental conditions, such as temperature and humidity, on RLPA performance.

- Extending the modeling framework to include deformable movers and more complex operational scenarios.

- Applying the proposed model to other piezoelectric actuator designs, such as rotary systems or multi-resonance actuators.

These efforts will further enhance the model's robustness and applicability, supporting the development of RLPA technologies for broader industrial and research applications.

## Acknowledgements

This work was supported by the Natural Science Research Start-up Foundation of Recruiting Talents of Nanjing University of Posts and Telecommunications, China (Grant No. NY223047).

## Conflict of interest

The authors declare no conflict of interest.

## Nomenclature

| | |
|---|---|
| $\delta$ | indentation depth of contact surface |
| $\eta$ | asperity density |
| $\mu_c$ | static friction coefficients |
| $\mu_s$ | sliding friction coefficients |
| $\mu_m$ | friction coefficients of guide rail |
| $\boldsymbol{L}$ | differential operator |
| $\nabla$ | gradient operator |
| $\omega$ | angular frequency |
| $\theta$ | phase difference |
| $\hat{\phi}$ | potential vector of element |
| $\Phi$ | global electric potential vector |
| $\psi$ | probability density of asperity height |
| $\rho$ | material density |
| $\rho_0$ | surface charge density |
| $\sigma$ | stress tensor |
| $\varepsilon$ | strain vector |
| $\varphi$ | electric potential |
| $\boldsymbol{c}$ | elastic stiffness constant matrix |
| $\mathbf{C}$ | Rayleigh damping matrix |
| $\mathbf{D}$ | electric displacement vector |
| $\mathbf{E}$ | electric field vector |
| $\boldsymbol{e}$ | piezoelectric stress constant matrix |
| $e_1$ | number of piezoelectric elements |
| $e_2$ | number of elastic elements |
| $E_0$ | shear modulus of the contact interface |
| $E_1$ | damping coefficient of the contact interface |
| $f$ | body force vector |
| $\mathbf{F}$ | surface force vector |
| $f$ | linear frequency |
| $\boldsymbol{F}^1$ | equivalent nodal force vector for the elastic region |
| $\boldsymbol{F}^2$ | equivalent nodal force vector for the piezoelectric region |
| $\mathbf{F}_n^e$ | equivalent element nodal force for normal contact force |
| $\boldsymbol{F}_{\text{pre}}$ | equivalent nodal force vector for pre-pressure |
| $\boldsymbol{F}_{\text{e}}$ | equivalent nodal force vector for electrical constraint |
| $f_{ti}$ | tangential nodal friction force |
| $f_{ni}$ | normal nodal friction force |
| $\mathbf{F}_n$ | global equivalent nodal force for normal contact force |
| $\boldsymbol{g}$ | permittivity constant matrix |
| $j$ | imaginary unit |
| $\boldsymbol{K}$ | global stiffness matrix |
| $\boldsymbol{K}_{uu}^1$ | mechanical stiffness matrix |
| $\boldsymbol{K}_{uu}^2$ | electrical stiffness matrix |
| $\boldsymbol{K}_{u\phi}$ | coupling matrix between mechanical and electric fields |

| | |
|---|---|
| $\boldsymbol{K}_{\phi\phi}$ | dielectric stiffness matrix for the piezoelectric region |
| $L$ | Lagrangian function |
| $\boldsymbol{L}$ | differential operator |
| $l$ | roughness |
| $\boldsymbol{M}$ | global mass matrix |
| $\boldsymbol{M}_{uu}^{1}$ | mass matrix for the elastic region |
| $\boldsymbol{M}_{uu}^{2}$ | mass matrix for the piezoelectric region |
| $m$ | mover mass |
| $\boldsymbol{N}$ | element shape functions matrix |
| $\mathbf{p}$ | element contact pressure vector |
| $q$ | body charge density |
| $\mathbf{Q}$ | equivalent charge force vector |
| $R$ | curvature radius of asperity |
| $T$ | kinetic energy |
| $U$ | amplitude of displacement |
| $u$ | time-domain displacement |
| $\mathbf{u}$ | displacement vector |
| $\hat{\boldsymbol{u}}$ | Nodal displacement vector of element |
| $U_E$ | electric energy in the electric field |
| $U_M$ | elastic strain energy |
| $W_M$ | work done by external mechanical forces |
| $W_E$ | work done by external electric forces |
| $z$ | asperity height |

## Author details

Ying He* and Ming Yuan
College of Automation and Artificial Intelligence, Nanjing University of Posts and Telecommunications, Nanjing, China

*Address all correspondence to: 20230032@njupt.edu.cn

## References


[1] Gao X, Yang J, Wu J, et al. Piezoelectric actuators and motors: Materials, designs, and applications. Advanced Materials Technologies. 2020; **5**(1):1900716. DOI: 10.1002/admt.201900716

[2] Mohith S, Upadhya AR, Navin KP, et al. Recent trends in piezoelectric actuators for precision motion and their applications: A review. Smart Materials and Structures. 2020;**30**(1):013002. DOI: 10.1088/1361-665X/abc6b9

[3] Mukhopadhyay S, Behera B, Kumar J. A brief review on the recent evolution in piezoelectric linear ultrasonic motors. Engineering Research Express. 2021; **3**(4):042003. DOI: 10.1088/2631-8695/ac3b73

[4] Zhao C. Ultrasonic Motors: Technologies and Applications. Berlin, Germany: Springer Science and Business Media; 2011

[5] Tian X, Liu Y, Deng J, et al. A review on piezoelectric ultrasonic motors for the past decade: Classification, operating principle, performance, and future work perspectives. Sensors and Actuators A: Physical. 2020;**306**:111971. DOI: 10.1016/j.sna.2020.111971

[6] Gao T, Liao Q, Si W, et al. From fundamentals to future challenges for flexible piezoelectric actuators. Cell Reports Physical Science. 2024;**5**(2): 101789. DOI: 10.1016/j.xcrp.2024.101789

[7] Kanchan M, Santhya M, Bhat R, et al. Application of modeling and control approaches of piezoelectric actuators: A review. Technologies. 2023;**11**(6):155. DOI: 10.3390/technologies11060155

[8] Wang Y, Deng J, Li H, et al. A resonant-type thin plate piezoelectric actuator inspired by koala's locomotion. IEEE Transactions on Industrial Electronics. 2023;**70**(8):8235-8243. DOI: 10.1016/j.ymssp.2020.107587

[9] Ling M, Zhang X. Coupled dynamic modeling of piezo-actuated compliant mechanisms subjected to external loads. Mechanism and Machine Theory. 2021; **160**:104283. DOI: 10.1016/j.mechmachtheory.2021.104283

[10] Yang L, Ren W, Ma C, et al. Mechanical simulation and contact analysis of the hybrid longitudinal-torsional ultrasonic motor. Ultrasonics. 2020;**100**:105982. DOI: 10.1016/j.ultras.2019.105982

[11] Liu Z, Wang H, Yang P, et al. Dynamic modeling and analysis of bundled linear ultrasonic motors with non-ideal driving. Ultrasonics. 2022;**124**: 106717. DOI: 10.1016/j.ultras.2022.106717

[12] Li Z, Zhu R, Yi X, et al. A high-power-density piezoelectric actuator operating in bicycling movement mechanism. IEEE Transactions on Industrial Electronics. 2022;**70**(6): 6090-6098. DOI: 10.1109/TIE.2022.3190900

[13] Zhang Y, Jin X, Yuan X, et al. Measurement and evaluation on mechanical performances of standing-wave piezoelectric motor of V-shaped sandwich-type transducer having arc tip. Measurement. 2022;**190**:110766. DOI: 10.1016/j.measurement.2022.110766

[14] DongMei X, BingJie Z, SiMiao Y, et al. Review on single-phase driven ultrasonic motors. Journal of Intelligent Material Systems and Structures. 2023; **34**(5):525-535. DOI: 10.1177/1045389X22112191

[15] Zhang JJ, Diao WD, Fan K, et al. A miniature standing wave linear

ultrasonic motor. Sensors and Actuators A: Physical. 2021;**332**:113113. DOI: 10.1016/j.sna.2021.113113

[16] Ryndzionek R, Sienkiewicz Ł. A review of recent advances in the single- and multi-degree-of-freedom ultrasonic piezoelectric motors. Ultrasonics. 2021; **116**:106471. DOI: 10.1016/j.ultras.2021.106471

[17] Naz S, Xu TB. A comprehensive review of piezoelectric ultrasonic motors: Classifications, characterization, fabrication, applications, and future challenges. Micromachines. 2024;**15**(9): 1170. DOI: 10.3390/mi15091170

[18] He J, Fan P, Ma X, et al. A fins-shaped bidirectional linear ultrasonic motor driven by a single-phase signal. Physica Scripta. 2024;**100**(1):015037. DOI: 10.1088/1402-4896/ad9e4d

[19] Lu D, Lin Q, Han Y, et al. A novel linear ultrasonic motor with dual sliders. AIP Advances. 2021;**11**(2):025238. DOI: 10.1063/5.0041085

[20] Wei W, Wu J, Ding Z, et al. A linear ultrasonic motor driven by torsional/bending vibrations. Sensors and Actuators A: Physical. 2023;**357**:114404. DOI: 10.1016/j.sna.2023.114404

[21] Lu D, Liu H, Xu J, et al. A linear ultrasonic motor based on coupling vibration mode. Micromachines. 2022; **13**(11):1852. DOI: 10.3390/mi13111852

[22] Askari M, Brusa E, Delprete C. Vibration energy harvesting via piezoelectric bimorph plates: An analytical model. Mechanics of Advanced Materials and Structures. 2023;**30**(23):4764-4785. DOI: 10.1080/15376494.2022.2104975

[23] Stefan JR. Piezoelectric Sensors and Actuators: Fundamentals and Applications. Berlin/Heidelberg, Germany: Springer; 2018

[24] Wang SY. A finite element model for the static and dynamic analysis of a piezoelectric bimorph. International Journal of Solids and Structures. 2004; **41**(15):4075-4096 j.ijsolstr.2004.02.058

[25] Zienkiewicz OC, Taylor RL. The Fnite Element Method for Solid and Structural Mechanics. Amsterdam: Elsevier; 2005

[26] Zhao T, Feng YT. Extended greenwood–Williamson models for rough spheres. Journal of Applied Mechanics. 2018;**85**(10):101007. DOI: 10.1115/1.4040537

[27] Marques F, Woliński Ł, Wojtyra M, et al. An investigation of a novel LuGre-based friction force model. Mechanism and Machine Theory. 2021;**166**:104493. DOI: 10.1016/j.mechmachtheory.2021.10449

[28] Deur J, Ivanović V, Troulis M, et al. Extensions of the LuGre Tyre friction model related to variable slip speed along the contact patch length. Vehicle System Dynamics. 2005;**43**(sup1):508-524. DOI: 10.1080/00423110500229808

[29] Nickell RE. Nonlinear dynamics by mode superposition. Computer Methods in Applied Mechanics and Engineering. 1976;**7**(1):107-129. DOI: 10.1016/0045-7825(76)90008-6

[30] He Y, Yao Z, Dai S, et al. Hybrid simulation for dynamic responses and performance estimation of linear ultrasonic motors. International Journal of Mechanical Sciences. 2019;**153**: 219-229. DOI: 10.1016/j.ijmecsci.2019.01.041

Chapter 2

# Vibration and Instability of Smart-Composite Sandwich Structure: Flutter and Divergences

*Zahra Khoddami Maraghi*

## Abstract

This study presents a comprehensive analysis of the vibration and instability characteristics of smart sandwich structures, focusing on beams, circular plates, and square plates with magnetorheological fluid (MRF) cores and magnetostrictive (Ms) skins. Timoshenko beam theory is applied to model the beam structures, accounting for both shear deformation and rotational inertia, while classical plate theory is used for the plate structures to capture the rheology behavior. The energy method is applied to derive the system's potential and kinetic energy, which are essential for evaluating the stability and vibration characteristics, and the differential quadrature method (DQM) is employed to solve the governing equations numerically. The research investigates the impact of magnetic fields on the natural frequency, loss factor, and system stability using several MRF models. Geometric parameters, including core and skin thickness, significantly affected the natural frequency and damping behavior, identifying flutter and divergences phenomena. The study also introduces a velocity feedback control parameter ($K_{vfc}$), which demonstrated an opposing effect on magnetic fields by enhancing damping and reducing the natural frequency. The research provides valuable insights for optimizing vibration control and stability in MRF-based sandwich structures.

**Keywords:** vibration, flutter and divergences, smart sandwich structures, MRF core, Ms. skins, velocity feedback control parameter

## 1. Introduction

Sandwich structures have gained attention for their excellent mechanical properties and versatility. Sandwich structures typically consist of two primary components: the core and the face sheets. The core provides shear resistance, energy absorption, and thermal insulation, while the face sheets offer high tensile and compressive strength. This synergistic configuration allows sandwich structures to bear bending loads while maintaining a low overall weight efficiently.

This study explores a novel class of sandwich structures incorporating MRF cores and magnetostrictive skins (MsS) or face sheets. This unique combination enhances vibration control and instability resistance, making these structures ideal for adaptive and smart applications in aerospace, mechanical, and civil engineering.

In recent years, several articles have investigated the vibrations and flutter instability of sandwich structures, often incorporating rheological fluids as the core. This section presents a concise review of some of these studies.

Nezami and Gholami [1] optimized the locations of MRF pockets in honeycomb sandwich beams for supersonic flutter suppression using a Genetic Algorithm. The study used classical beam theory, Winkler-Pasternak foundation, and supersonic piston theory. Optimized MRF pocket placement enhanced vibrational characteristics and postponed flutter to higher dynamic pressures. The research identified the best number of MR pockets and magnetic field values for optimal flutter suppression. Ghorbanpour Arani and Soleymani [2] studied the size-dependent vibration behavior of a sandwich beam with an MR core and functionally graded material (FGM) face layers under yawed supersonic airflow. They applied the linear piston theory for aerodynamic pressure and the modified first strain gradient theory (MFSGT) for size-dependent strain energy. Their results explored the effects of aerodynamic pressure, axial speed, magnetic field intensity, yaw angle, and power-law index on natural frequencies and loss factors, providing insights into the critical aerodynamic pressure and axial speed under various conditions. Khoddami Maraghi [3] investigated the dynamic instability (flutter) of sandwich composite plates under structural deformations and follower forces. Effective elastic properties of CNT-reinforced composites are derived using the rule of mixture, with Pasternak foundation modeling shear and normal modulus. Terfenol-D layers are employed for feedback control, influencing flutter behavior. Results highlight critical follower forces for flutter onset, considering thickness ratios, feedback gain, fiber volume fraction, and temperature gradients. Aboutalebi et al. [4] investigated the nonlinear vibration behavior of circular, annular, and sector sandwich panels with MRF cores. The MRF operated in the post-yield region, affecting vibration attenuation. Using finite element analysis, the study explored the impact of magnetic fields and maximum deformation on resonant frequencies and loss factors, emphasizing the role of MR fluid's post-yield behavior in dynamic performance. Sharif et al. [5] reviewed the use of MR elastomers (MREs) in sandwich structures for vibration control. They explored the field-dependent stiffness and damping properties of MREs, analyzing their impact on free and forced vibrations, stability, and optimal placement in structures. The review highlights key research on MRE-based structures for improved vibration suppression and stability in various applications. Kuriakose and Sreehari [6] experimentally investigated the use of piezoelectric patches to enhance the vibration and flutter characteristics of damaged composite plates. Internal flaws reduced the natural frequency, but piezoelectric patches restored it. Vibration amplitude was controlled by up to 30% with applied voltage. Wind tunnel tests showed improved flutter velocity when voltage was applied to the piezoelectric patches on damaged plates. Amiri et al. [7] developed a model for the dynamics and stability of sandwich micro-shells with an MR elastomer core subjected to moving fluid flow. Using Donnell's shell theory and MFSGT, they analyzed the system's dynamic characteristics and stability. Their findings showed the MRE core's significant impact on reducing instabilities and improving system stability. Sarafraz et al. [8] studied the flutter behavior of a sandwich plate with an auxetic honeycomb core and composite face sheets reinforced with graphene nanoplatelets and fibers under yawed supersonic flow. They used sinusoidal shear deformation theory and linear piston theory, solving the equations *via* DQM. Results showed that increasing the auxetic core thickness reduced critical aerodynamic pressure, while core geometry had minimal effect. Omidi Soroor et al. [9] studied the vibration behavior of sandwich beams with axially graded face sheets and an MR core. They used Euler-Bernoulli and Timoshenko beam theories along with the Rayleigh-Ritz method to analyze the system's dynamic properties. Their results showed that material distribution, gradient index, boundary conditions, core thickness, and magnetic field significantly affect the vibration frequency and loss factor. Oveissi et al. [10] studied energy harvesting from vortex-induced vibrations of an axially moving composite nanoshell conveying nanofluid. The nanoshell consists of piezoelectric, MR elastomer, and CNT layers. They found that decreasing fluid densities and adjusting the magnetic field intensity significantly enhanced the output voltage. The study highlights the potential of using vortex-induced vibrations for renewable energy harvesting in nanostructures. Wan et al. [11] examined the nonlinear flutter response and reliability of smart hybrid nanocomposite trapezoidal plates under supersonic yawed flow conditions. It integrated carbon nanotubes (CNTs) and carbon fibers, actuators, and sensors for enhanced performance. The research explored the effects of material properties, geometry, and control mechanisms. Findings show significant improvements in flutter critical pressure and frequency, with reliability enhancements through CNTs and voltage adjustments. Moreira et al. [12] studied the layerwise finite element modeling and active aeroelastic flutter control of smart variable stiffness composite panels with surface-bonded piezoelectric layers under supersonic airflow. The models combined shear deformation theories and electric potential distribution with supersonic airflow effects. The research explored the impact of proportional control on flutter response with various airflow directions and piezoelectric patch placements. Jonna and Srinivas [13] studied the dynamic behavior and aeroelastic stability of rotating tapered composite sandwich blades with an MR elastomer core under supersonic flow. Using Timoshenko beam theory and finite element methods, they examined the effects of magnetic field intensity, taper ratio, MR layer thickness, and rotating speed on critical aerodynamic pressure and damping. A neural network regression model was developed to optimize these parameters.

Sandwich structures with MRF cores and magnetostrictive materials (MsMs) face sheets represent a breakthrough in adaptive structural design, addressing challenges in vibration and instability under dynamic conditions. The MRF core, whose viscosity can be tuned *via* a magnetic field, transitions rapidly between liquid-like and semi-solid states, providing dynamic control over mechanical properties. Similarly, MsMs, such as Terfenol-D, alter stiffness in response to magnetic fields, enabling real-time vibration suppression and load adaptability. These interactions, modeled by governing equations accounting for core shear deformation and magnetic field effects, allow precise control of vibration and instability characteristics. This innovation has significant applications in aerospace panels, vibration-damping bridge decks, and smart building facades, where enhanced performance, safety, and efficiency are crucial. Future research will focus on optimizing magnetic field strategies and ensuring the long-term reliability of these adaptive systems.

## 2. Structural definition

A schematic figure of a sandwich structure with three layers is illustrated in **Figure 1**, in which geometrical parameters of length $(l, a)$, width $(b)$, skin thickness $(h_t)$, and core thickness $(h_c)$ along with the coordinate axis are indicated.

As shown in **Figure 1**, the structure analyzed in this paper consists of three layers:

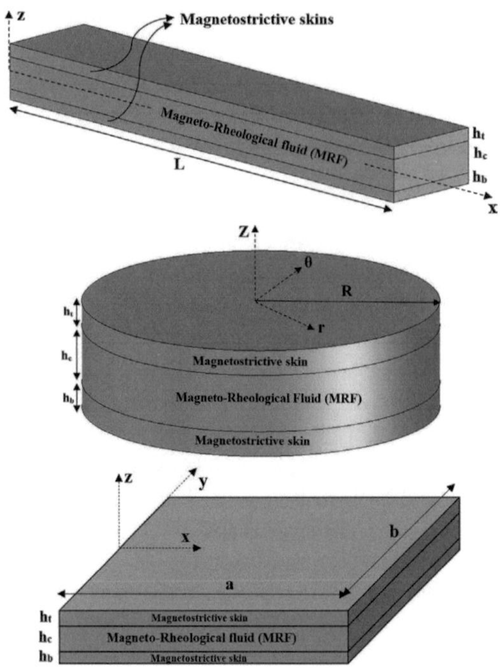


**Figure 1.**
*A schematic figure of a sandwich structure.*

- A central layer filled with MRF,
- Two surface layers made of MsM.

The strain-displacement relations are derived separately for each layer, followed by the calculation of forces and moments. Finally, the total energy equation, incorporating the contributions from both the MR fluid core and the MsSs, is formulated using Hamilton's principle.

## 3. Constitutive equations

### 3.1 Structure

#### *3.1.1 Beam*

Various beam theories related to deformation kinematics are currently used in structural analysis. To explain these beam theories, it is necessary to first describe the coordinate system. The analysis considers three axes: the longitudinal axis of the beam, the thickness direction (maximum thickness), and the transverse axis of the beam. In general, in beam theory, all applied loads, geometry, and displacements are functions of these coordinates. It is typically assumed that displacements along certain directions are negligible. In this study, Timoshenko beam theory is applied for the analysis [14]:

$$\begin{aligned} u(x,z,t) &= u_0(x,t) - z\varphi_0(x,t), \\ w(x,z,t) &= w_0(x,t) \end{aligned} \tag{1}$$

w0 and u0 are the in-plane and vertical displacement as shown **Figure 1** and $\varphi_0$ is in-plane rotation, respectively.

*3.1.2 Plate*

The total displacement components at a material point (x,y,z) for classical plate theory can be expressed as follows [15, 16]:
Rectangle plate

$$\begin{aligned} U_{1i}(x,y,z,t) &= u_i(x,y,z) - z\frac{\partial w}{\partial x}, \\ U_{2i}(x,y,z,t) &= v_i(x,y,z) - z\frac{\partial w}{\partial x}, \\ U_{3i}(x,y,z,t) &= w(x,y), \end{aligned}$$

Circular plate

$$\begin{aligned} U_{1i}(r,\theta,z,t) &= u_i(r,\theta,t) - z\frac{\partial w}{\partial r}, \\ U_{2i}(r,\theta,z,t) &= v_i(r,\theta,t) - z\frac{\partial w}{\partial r}, \\ U_3(r,\theta,z,t) &= w(r,\theta,t) \end{aligned} \tag{2}$$

In this context, U1i and U2i ($i$ = t, b), represent the displacements of an arbitrary point on the skins in cylindrical and rectangular coordinates, respectively, while U3 denotes the displacement along the thickness direction.

*Kinematics of MRF*: MRF behave like Newtonian fluids when no magnetic field is applied. However, in the presence of a magnetic field, their behavior changes, and they act like a solid gel. The yield point is critical in determining the rheological response of these materials. For the MR core, the following assumptions are made [17]:

- The MRF is isotropic.
- The MRF core behaves as a Bingham plastic material.
- The normal stress in the MRF core is neglected because Young's modulus of the MRF is negligibly small compared to that of the MsSs.
- There is no slip between the layers.
- The transverse displacement is uniform across all points of a cross-section.

Based on these assumptions, the strain-displacement relations for the MRF are given as follows [18]:

$$\varepsilon_{ij} = 0.5\left[\nabla U_k + (\nabla U_k)^T\right], k = 1,2,3 \tag{3}$$

where εij (i, j = r, θ, z) are the strain components. Using Eq. (3) for MRF [19]:

$$\gamma_{xz2} = \frac{h_t + 2h_c + h_b}{2h_c}\frac{\partial w}{\partial x} + \frac{u_t - u_b}{h_c},$$

$$\gamma_{yz2} = \frac{h_t + 2h_c + h_b}{2h_c}\frac{\partial w}{\partial y} + \frac{v_t - v_b}{h_c},$$

$$\gamma_{rz}^c = \left[\frac{1}{2h_c}(h_t + 2h_c + h_b)\right]\frac{\partial w}{\partial r} + \frac{(u_t - u_b)}{h_c},$$
$$\gamma_{\theta z}^c = \left[\frac{1}{2h_c}(h_t + 2h_c + h_b)\right]\frac{\partial w}{r\partial\theta} + \frac{(v_t - v_b)}{h_c}, \tag{4}$$

Based on linear viscoelastic theory, the shear stress and strain are related in the pre-yield region as follows [20]:

$$\{\tau_{rz},\ \tau_{\theta z}\} = G_c\{\gamma_{rz}, \gamma_{\theta z}\} \tag{5}$$

where $G_c$ represents the complex shear modulus of the MRF, characterizing its viscoelastic behavior in the pre-yield regime.

### 3.2 MR fluid

MRFs are smart materials composed of magnetic particles suspended in a carrier fluid. When exposed to an external magnetic field, the particles align, rapidly changing the fluid's viscosity and mechanical properties, making MRFs useful for adaptive damping and vibration control systems. The complex shear modulus ($G$) is provided because the Bingham plastic model does not mostly present a detailed and accurate definition of the behavior of MRF. Complex shear modulus is divided into two parts: real modulus (storage modulus) and imaginary modulus (loss modulus). For MRF, this property is dependent on the external magnetic field [21]. The storage modulus is proportional to the average energy stored during a cycle of deformation per unit volume of MRF, and the loss modulus is proportional to the average energy dissipated per unit volume of MRF over a cycle.

To study the effect of different types of MRFs, three types of them are considered. The second type of MRFs was presented by Ramamoorthy et al. [22]:

$$G' = -0.05035B^2 + 428.455B + 858.8,$$
$$G'' = -0.057B^2 + 452.105B + 848.35,$$
$$\rho = 2812 kg/m^3 \tag{6}$$

Aguib et al. [23] introduced the third type of MRF with the following specifications:

$$G' = -3.238 \times 10^{-6}B^3 + 0.02733B^2 + 38.29B + 1600000,$$
$$G'' = -6.889 \times 10^{-6}B^3 + 0.02122B^2 + 70.11B + 330000,$$
$$\rho = 1100\ kg/m^3 \tag{7}$$

Note that the aforementioned specifications are gained based on experimental tests, and B symbolizes the magnetic field intensity in Gauss. The following higher-order field describes how the mechanical properties of an MRF core change with an applied magnetic field B in Tesla (T) [24]:

$$\begin{aligned}
&G_2 = G_0(1 + i\eta_0)\\
&G_0 = 2.7006 + 2.43B + 38.778B^2 - 93.422B^3 + 71.797B^4 - 9.1077B^5 - 6.9395B^6 \ MPa,\\
&\eta_0 = 0.228 + 0.1526B + 2.3522B^2 - 12.185B^3 + 22.148B^4 - 17.787B^5 + 5.3485B^6,\\
&\rho = 3500\ kg/m^3
\end{aligned} \tag{8}$$

*Ramamoorthy et al. [22]*: Accurate and practical for moderate field intensities but limited for high-intensity applications.

*Aguib et al. [23]*: Extended range and improved accuracy for higher-intensity fields, with broader applicability in demanding scenarios.

*Higher-order field model*: Most accurate and versatile, suitable for extreme conditions but computationally intensive.

### 3.3 Magnetomechanical coupling effect

MsM exhibit *magnetic anisotropy* in their atomic structure, causing them to undergo dimensional changes when subjected to magnetic fields. To account for the effect of magnetic fields on MsM, the *magnetomechanical coupling* for isotropic MsM can be represented by the following matrices.

#### *3.3.1 Beam*

The layers of this structure consist of the background matrix and nanoplatelets made of graphene. Once the strain is calculated, the corresponding stresses can be derived, and the relationship between strain and stress, according to *Hooke's law*, is expressed as follows. In this equation, the vertical stress in the Z direction is neglected, effectively simplifying it into a two-dimensional matrix [25]:

$$\begin{aligned}
&\begin{bmatrix}\sigma_{xx}\\ \tau_{xz}\end{bmatrix} = \begin{pmatrix}C_{11} & 0\\ 0 & C_{55}\end{pmatrix}\begin{bmatrix}\varepsilon_{xx}\\ \gamma_{xz}\end{bmatrix} - \begin{pmatrix}0 & e_{31}\\ 0 & e_{35}\end{pmatrix}\begin{bmatrix}0\\ H_z\end{bmatrix}, \ C_{11} = \frac{E}{1-\nu^2}, C_{55} = \frac{E}{2(1+\nu)}\\
&e_{31} = \bar{e}_{31}\cos^2\theta + \bar{e}_{35}\sin^2\theta,\\
&e_{35} = \bar{e}_{31}\sin^2\theta + \bar{e}_{35}\cos^2\theta,
\end{aligned} \tag{9}$$

where $\sigma_{ij}$ and $\varepsilon_{ij}$ refer to the *magnetic stress* and *strain*, respectively, and E, G, and $\nu$ are the Yong modulus, shear modulus, and Poisson's ratio, respectively.

$H_z$ and $e_{ij}$ are the magnetic field intensity and the Ms. coupling coefficients, respectively, which can be expressed as follows [26–28]:

$$H_z = K_{vfc}\frac{\partial w_0(x,y,t)}{\partial t} \tag{10}$$

where $K_{vfc}$ $(A.s/m^2)$ is introduced as velocity feedback control parameter. Where $\theta$ represents the direction of magnetic anisotropy and the values of $\bar{e}_{ij}$ are reported in $\bar{e}_{ij} = 442.55\ (N/(m.A))$ where E = 30 GPa, $\rho$ = 9250 kg/m$^3$, and $\nu$ = 0.25.

### 3.3.2 Circular plate

The constitutive relations for each of the MsS may be established as [29]:

$$\begin{Bmatrix} \sigma_{rr} \\ \sigma_{\theta\theta} \\ \sigma_{r\theta} \end{Bmatrix} = \begin{bmatrix} C_{11} & C_{12} & 0 \\ C_{12} & C_{22} & 0 \\ 0 & 0 & C_{66} \end{bmatrix} \begin{Bmatrix} \varepsilon_{rr} \\ \varepsilon_{\theta\theta} \\ \gamma_{r\theta} \end{Bmatrix} - \begin{bmatrix} 0 & 0 & e_{31} \\ 0 & 0 & e_{32} \\ 0 & 0 & e_{36} \end{bmatrix} \begin{Bmatrix} 0 \\ 0 \\ H_z \end{Bmatrix}$$
$$C_{11} = C_{22} = \frac{E}{(1-\nu^2)}, \quad C_{12} = \frac{E\upsilon}{(1-\nu^2)}, C_{66} = \frac{E}{2(1+\nu)} \tag{11}$$

where $\sigma_{ij}$ and $\varepsilon_{ij}$ are stress and strain tensors, respectively.

### 3.3.3 Rectangular plate

The relationship between strain and stress in the plate, according to Hooke's law [30]:

$$\begin{bmatrix} \sigma_{xx} \\ \sigma_{yy} \\ \sigma_{xy} \end{bmatrix} = \begin{bmatrix} C_{11} & C_{12} & 0 \\ C_{21} & C_{22} & 0 \\ 0 & 0 & C_{66} \end{bmatrix} \begin{bmatrix} \varepsilon_{xx} \\ \varepsilon_{yy} \\ \varepsilon_{xy} \end{bmatrix} - \begin{bmatrix} 0 & 0 & e_{31} \\ 0 & 0 & e_{32} \\ 0 & 0 & e_{36} \end{bmatrix} \begin{bmatrix} 0 \\ 0 \\ H_z \end{bmatrix}$$
$$C_{11} = C_{22} = \frac{E}{(1-\upsilon^2)}, \quad C_{12} = \frac{E\upsilon}{(1-\upsilon^2)}, C_{66} = \frac{E}{2(1+\upsilon)} \tag{12}$$

where $\sigma_{ij}$ and $\varepsilon_{ij}$ refer to the *magnetic stress* and *strain*, respectively, which are influenced by $H_z$, the magnetic field intensity and *Ms coupling coefficients* $e_{ij}$ can be expressed as follows [26–28]:

$$\begin{aligned} e_{31} &= \bar{e}_{31} \cos^2\theta + \bar{e}_{32} \sin^2\theta, \\ e_{32} &= \bar{e}_{31} \sin^2\theta + \bar{e}_{32} \cos^2\theta, \\ e_{36} &= (\bar{e}_{31} - \bar{e}_{32}) \sin\theta \sin\theta. \end{aligned} \tag{13}$$

The values of $\bar{e}_{ij}$ are reported in Section 3.3.1.

## 4. Energy method in sandwich plate

### 4.1 Strain energy

The *energy method* is employed to derive the governing equations. The strain energy of the rectangular sandwich plate is calculated as follows [31]:

$$\begin{aligned} U_{Sandwich} &= U_{MRFcore} + U_{MsSkin}^{Bottom} + U_{MsSkin}^{Top} \\ U_{MRFcore} &= \int_V G\left(\gamma_{xz}^2 + \gamma_{yz}^2\right) dV \\ U_{MsSkin} &= \frac{1}{2} \int_V \left(\sigma_{ij}\varepsilon_{ij}\right) dV \end{aligned} \tag{14}$$

### 4.2 Kinetic energy

The kinetic energy of sandwich plate is composed of three parts $\left(K_{Sandwich} = K_{MRFcore} + K_{MsSkin}^{Bottom} + K_{MsSkin}^{Top}\right)$:

$$K_{Skins} = \frac{1}{2}\int_V \left[\rho_i\left(\left(\frac{\partial u_i}{\partial t}\right)^2 + \left(\frac{\partial v_i}{\partial t}\right)^2\right)\right] dV, i = t, b \tag{15}$$

Kinetic energy corresponding to transverse displacement:

$$K_{MRF\,core,MsSkins} = \frac{1}{2}\int_V \left[\rho_i\left(\frac{\partial w}{\partial t}\right)^2\right] dV \tag{16}$$

Kinetic energy corresponding with the rotation of MRF core:

$$K_{MRF\,core} = \frac{1}{2}\int_A \rho z\left[\left(\frac{\partial \gamma_{xz}}{\partial t}\right)^2 + \left(\frac{\partial \gamma_{yz}}{\partial t}\right)^2\right] dA \tag{17}$$

*Hamilton's principle* is used to derive the equations of motion. The principle can be stated analytically, where the first variation of the motion equations must be zero, as follows [31]:

$$\delta\int_{t_1}^{t_2}[U_{Sandwich} - K_{Sandwich} - W_F]dt = 0 \tag{18}$$

where $\delta U_{Sandwich}$ and $\delta T_{Sandwich}$ represent the variation of strain energy and the variation of kinetic energy, respectively. Also, $\delta W_F$ is the work done by external forces as $F=F_0(x\text{-}a)$ for rectangular plate and beam and $F=F_0(r\text{-}r_0)$ for circular plate $F_0$(N).

### 4.3 Solution procedure

Initially, Eq. (19) is employed to separate the variables into spatial and temporal components [32]:

$$\Psi(x,y,t) = \Psi(x,y)\ e^{\omega t}, \Psi \to U_1, U_2, V_1, V_2, W, \varphi \tag{19}$$

Here, ω (*rad/s*) represents the natural frequency. In the differential quadrature method (DQM), the differential equations are transformed into algebraic equations, where the partial derivatives of a function (F) are approximated using specific weighting coefficients at discrete grid points. The function F is assumed to represent U, V, W, and ϕ with respect to the spatial variables x and y ($0 < x < L, 0 < y < L$). The grid points along these variables are denoted by $Nx \times Ny$. According to Ref. [33], the derivative of F is expressed as:

$$\frac{d^n F\left(x_i, y_j\right)}{dx^n} = \sum_{k=1}^{N_x} A_{ik}^{(n)} F\left(x_k, y_j\right) \qquad n = 1, \dots, N_x - 1,$$

$$\frac{d^m F\left(x_i, y_j\right)}{dy^m} = \sum_{l=1}^{N_y} B_{jl}^{(m)} F(x_i, y_l) \qquad m = 1, \dots, N_y - 1,$$

$$\frac{d^{n+m} F\left(x_i, y_j\right)}{dx^n dy^m} = \sum_{k=1}^{N_x} \sum_{l=1}^{N_y} A_{ik}^{(n)} B_{jl}^{(m)} F(x_k, y_l) \tag{20}$$

where $A_{ik}^{(n)}$ and $B_{jl}^{(n)}$ are the weighting coefficients determined using Chebyshev polynomials for the positions of the grid points, with their recursive formulas provided in Shu [33]. By applying DQM and substituting Eq. (19) into the governing equations, the standard form of the vibrational motion equation set $\left(M\partial^2\psi/\partial t^2 + C\partial\psi/\partial t + K\psi = 0\right)$ is derived. Considering simply supported boundary conditions, an eigenvalue problem is obtained, where the eigenvalues of the state-space matrix correspond to the dimensionless frequencies. It is worth noting that {M} is the mass matrix, {C} is the damping matrix, {K} is the stiffness matrix, and {I} and {0} represent the unit and zero matrices, respectively.

The results in this section are presented with a focus on two dynamic parameters: natural frequency ω (*rad/s*) and loss factor. Due to the presence of damping, the natural frequency is expressed as a complex value, and the ratio of its real part to its imaginary part defines the loss factor $(Re(\omega)/Im(\omega))$, yielding a dimensionless expression.

## 5. Numerical results and discussion

This section discusses the results of vibration and instability analysis for three-layer sandwich structures, specifically focusing on beams, circular plates, and square plates. The sandwich structures consist of MsM material for the surface layers and MRF for the core. The surface material properties are outlined as follows, and the core material properties are detailed subsequently.

The results are categorized into three groups:

- uniform beam,
- circular plate, and
- square plate.

Care has been taken to select models that allow for comparative analysis between these structures. However, due to the unique behavior of MRF, it is not possible to achieve identical conditions across all three structures. In this chapter, we introduced three models of MRF, each tested across the three structures. The best extraction results for each case are presented.

The results presented in **Table 1** validate the accuracy of the current formulation by comparing the natural frequencies with those reported by Nayak et al. [34] and

| Natural frequencies (ω) | Nayak et al. [34] | Asghari [35] | Present analysis |
|---|---|---|---|
| First mode | 57.146 | 57.139 | 57.14 |
| Second mode | 223.919 | 219.585 | 222.303 |
| Third mode | 465.932 | 465.171 | 465.873 |

**Table 1.**
*Comparison of the first three natural frequencies of the MR sandwich beam under simply supported boundary conditions.*

Asghari et al. [35] under simply supported boundary conditions. The comparison demonstrates excellent agreement, confirming the reliability and precision of the proposed method in predicting the vibrational characteristics of the structure.

### 5.1 Beam

As previously described, the MRF was formed with the ability to respond to magnetic fields. When exposed to magnetic fields of specific strengths, magnetic dipoles form within the fluid, transitioning it from a liquid to a solid-like gel state. This transition naturally increases the material's stiffness, which in turn raises its vibration frequency. It is worth noting that **Figure 2** is for Ramamoorthy et al. [22] model. In the working model, low magnetic fields also effectively increase the natural frequency. As shown in **Figure 2**, a noticeable change in the natural frequency occurs when the magnetic field increases to 100 Gauss. Predictions indicate that applying a magnetic field will increase the natural vibration frequency of MRF, and **Figure 2** supports this. As shown, the natural frequency increases as the magnetic field strength increases. This trend is observed consistently across the first three vibration modes. Conversely, the loss factor (damping coefficient) initially shows a brief increase but then decreases sharply as the magnetic field increases. This outcome is logical: as the

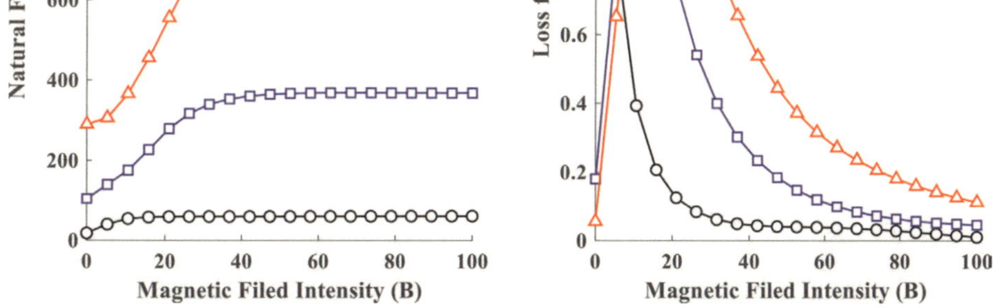


**Figure 2.**
*Variation of natural frequency and loss factor of sandwich beam versus magnetic field intensity in three modes for Ramamoorthy et al. [22].*

MRF becomes stiffer, its damping ability reduces, leading to a decrease in the coupling coefficient.

**Figure 3** presents results derived from Aguib et al. [23] model. In this model, to observe meaningful changes, the magnetic field intensity needs to be higher than in the Ramamoorthy et al. [22] model. Therefore, the magnetic field is increased up to 4000 Gauss.

As with previous observations, the natural frequency of vibrations in the first three modes increases with an increase in the magnetic field. This trend occurs because the MFR stiffens under the magnetic field, naturally raising the frequency. However, the loss factor (damping coefficient) decreases in subsequent vibration modes, following a trend distinct from the surface model. If the magnetic field exceeds 4000 Gauss, the results become undesirable due to an unexpected increase in the damping factor. Therefore, in this model, the magnetic field intensity is limited to 4000 Gauss to maintain reliable results.

**Figure 4** illustrates the changes in natural frequency and loss factor for three vibration modes in the high-order model. Similar to the previous two models, increasing the magnetic field results in a noticeable rise in natural frequency and a subsequent decrease in the loss factor. However, this model requires a higher magnetic field intensity, unlike the earlier models. To achieve the desired results, the magnetic field has been increased to 1 Tesla.

Among the three models, Ramamoorthy et al. [22] model is selected for further investigation. **Figures 5** and **6** illustrate the effects of geometric parameter variations on the natural frequency and instability of the sandwich beam.

**Figure 5** shows the effect of increasing core thickness on the natural frequency and loss factor. As seen in **Figure 5**, increasing the core thickness leads to a rise in natural frequency. This occurs because when the MRF core occupies a larger portion of the structure and is exposed to the magnetic field, the overall stiffness increases.

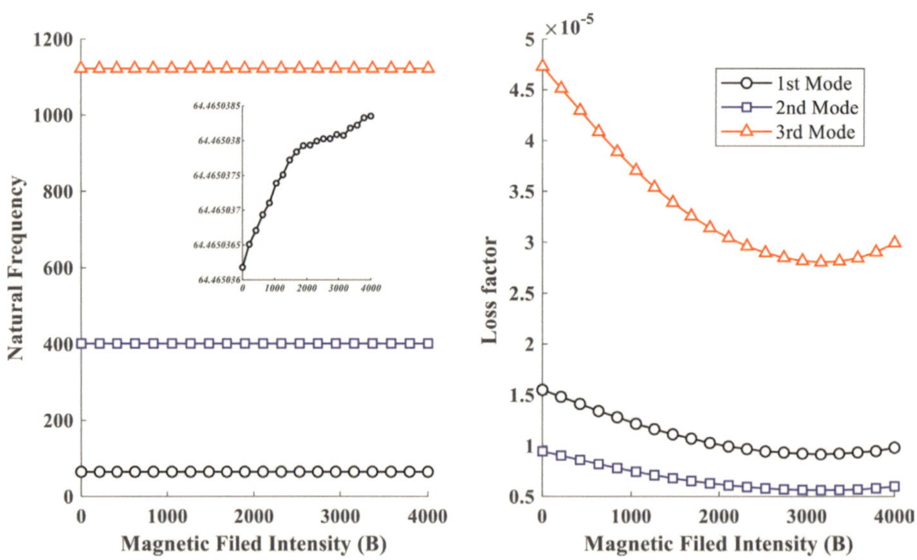


**Figure 3.**
*Variation of natural frequency and loss factor of sandwich beam versus magnetic field intensity in three modes for Aguib et al. [23].*

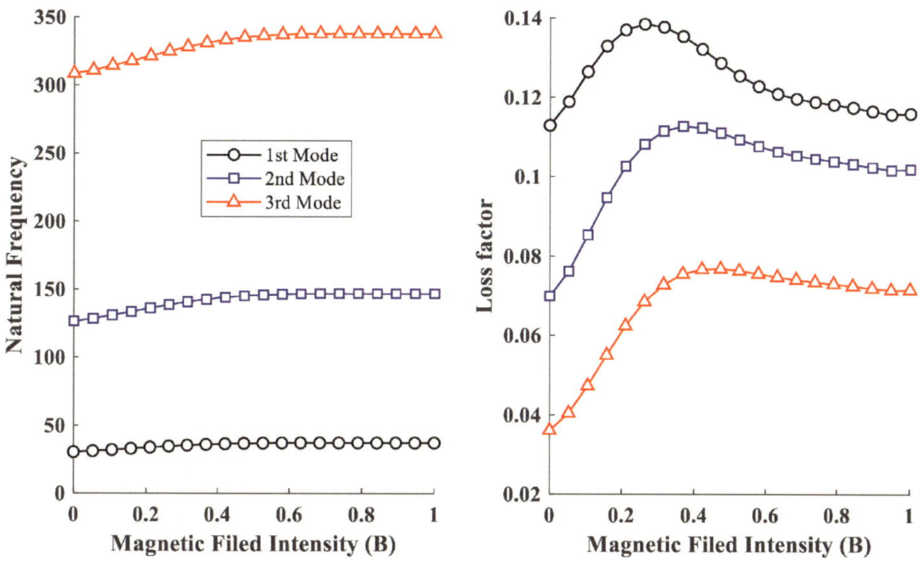


**Figure 4.**
*Variation of natural frequency and loss factor of sandwich beam versus magnetic field intensity in three modes for high-order model.*

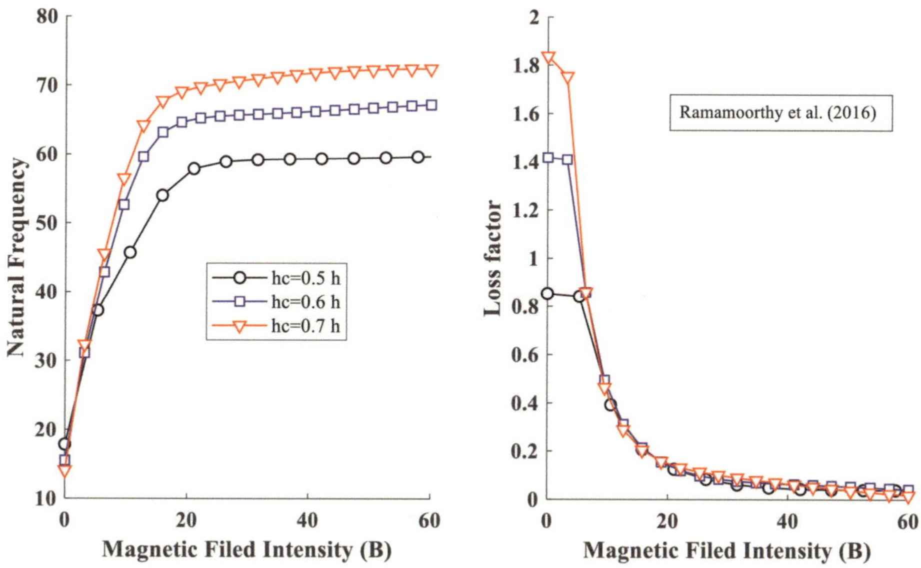


**Figure 5.**
*Variation of natural frequency and loss factor of sandwich beam versus magnetic field intensity in different thicknesses of core.*

In this analysis, the core thickness is set to 0.5, 0.6, and 0.7 of the total sandwich beam thickness. As these values increase, the natural frequency also increases.

It is important to note that the total thickness of the sandwich beam remains constant as the MRF core thickness increases. Therefore, an increase in core thickness results in a decrease in the thickness of the MsM skins. **Figure 5** illustrates the results

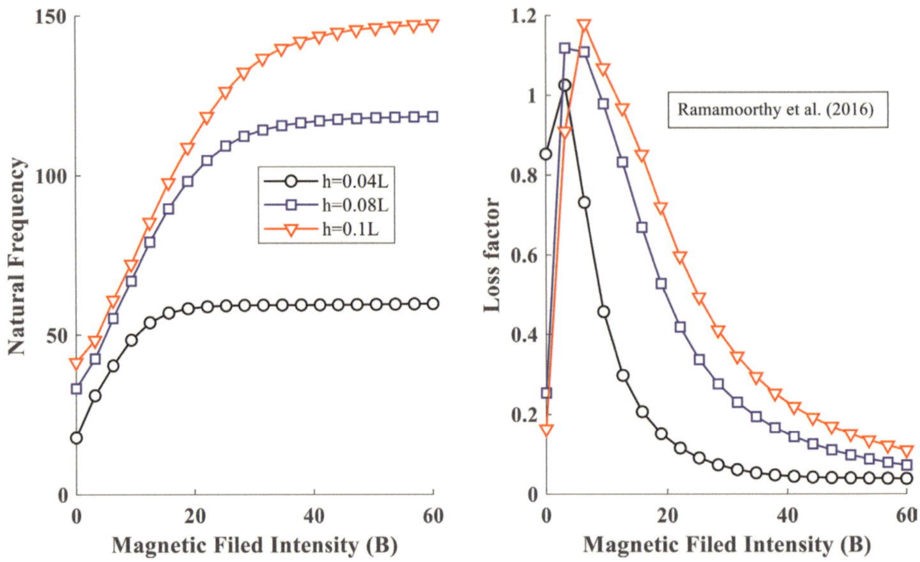


**Figure 6.**
*Variation of natural frequency and loss factor of sandwich beam versus magnetic field intensity in different total thicknesses of sandwich beam.*

when the MRF core occupies a larger proportion of the sandwich beam than the MsM skins.

**Figure 6** shows how the natural frequency of the sandwich structure changes as the total thickness increases. In this case, both the MRF core and the MSM layers contribute to the overall thickness. As seen in **Figure 6**, increasing the core thickness under a given magnetic field raises the natural frequency, as both the MRF fluid and the MSM layers enhance the structure's stiffness. Additionally, the increase in natural frequency leads to a reduction in the loss factor.

A non-conservative force is a force for which the work done depends on the path taken and not just on the initial and final positions. Unlike conservative forces, non-conservative forces cause energy dissipation, often in the form of heat or sound. The work done by non-conservative forces cannot be fully recovered, as some energy is lost to the surroundings. Examples of non-conservative forces include friction, air resistance, and viscous damping. In systems with non-conservative forces, mechanical energy (kinetic + potential energy) is not conserved because energy is lost to other forms like heat or sound. When conservative forces are applied to a system, its vibrations are accompanied by damping. Aerodynamic forces are among these conservative forces. In this chapter, we discuss the forces involved in considering the effect on stability. Unsteadiness arises when the system passes critical values, and the overlap of vibration modes—specifically the first and second modes—can lead to flutter. Flutter is an instability phenomenon that occurs in structures, particularly in aerodynamics and mechanical systems, when certain vibration modes interact with non-conservatives. It typically happens when these forces cause the system to oscillate with increasing amplitude, leading to rapid and uncontrollable vibrations. Flutter usually occurs when the system's natural frequency coincides with the frequency of an external force (such as airflow) and when vibration modes, like the first and second modes, overlap. This can result in structural damage or failure if not controlled.

The unstable point in **Figures 7** and **8** is where the right-hand graph crosses the dotted line and goes negative; this point indicates instability. The higher this point is, the greater the critical force, and thus, the better the stability. In other words, the structure can withstand higher force without becoming unstable. As this value increases, the stiffness of the structure also increases. In this context, the mass of the structure has a lesser effect on the process, although it still influences the frequency.

Divergence is a static instability where a structure, such as a beam, experiences a continuous deformation under aerodynamic loading $F_0$(N) until failure. It occurs when aerodynamic moments or forces overcome the elastic restoring forces, causing the structure to bend or twist uncontrollably. Unlike flutter, divergence does not involve oscillations. In **Figures 7** and **8** on the right, the point where the curves begin to bifurcate indicates the onset of divergence-type instability.

## 5.2 Circular plate

**Figures 9** and **10** illustrate the changes in natural frequency and the loss factor for Ramamoorthy et al. [22] and Aguib et al. [23] models. As noted for the sandwich beam, the circular sandwich plate with an MRF core and MsM surface shows a consistent trend: increasing the magnetic field raises the natural frequency and significantly reduces the loss factor. This occurs because the MRF core stiffens under the magnetic field, increasing the frequency while reducing the system's damping due to MRF excitation.

The magnetic field strength differs between the two models. In Ramamoorthy et al. [22], the magnetic field for the circular plate is evaluated up to 6000 Gauss, while in Aguib et al. [23] model, it is applied up to 7000 Gauss.

In the vibration and instability analysis of a circular sandwich plate, a model is selected to explore additional parameters, including geometric variations. As shown in

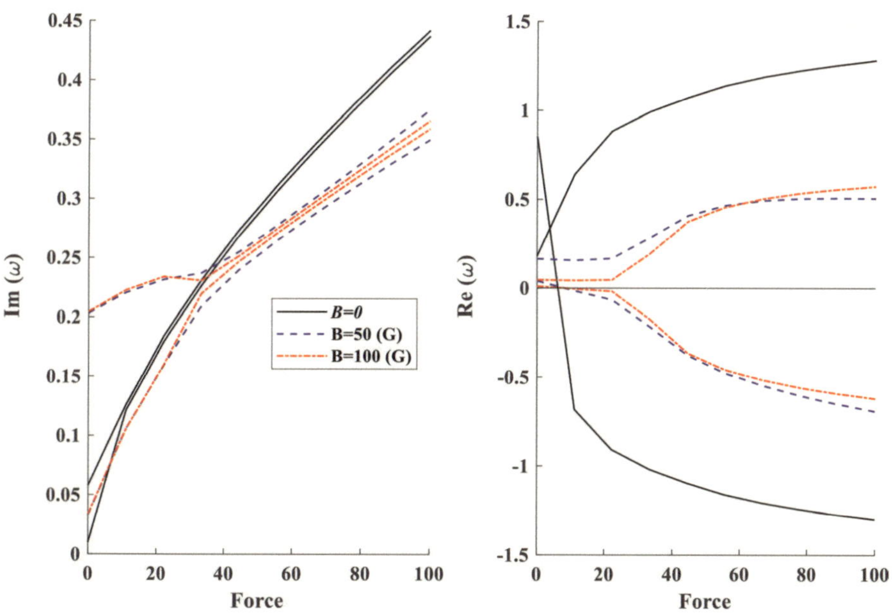


**Figure 7.**
*Variation of real and imagery of frequency versus non-conservative force in different magnetic field intensity of sandwich beam for Ramamoorthy et al. [22].*

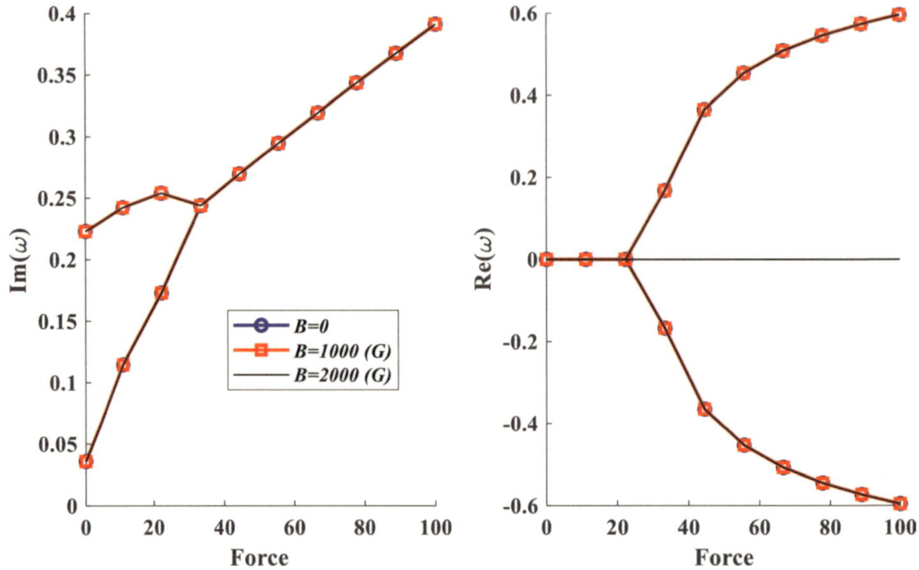


**Figure 8.**
*Variation of real and imagery of frequency versus non-conservative force in different magnetic field intensity of sandwich beam for Aguib et al. [23].*

**Figure 11**, increasing the core thickness hc(m) leads to a rise in the natural frequency of vibrations and a sharp decrease in the coupling coefficient. This occurs because increasing the volume of the MR fluid enhances the system's stiffness when subjected to an appropriate magnetic field.

It is worth noting that in all these cases, the thickness of the magnetic layers is kept constant. This approach is reasonable for maintaining consistency in the analysis. Another point that can be extracted from **Figure 11** is the increase in the loss factor after a magnetic field with an intensity of 3500 Gauss. It is important to note that the magnetic field intensity applied to the MRF fluid has an upper limit; beyond this limit, the behavior of the fluid changes significantly.

**Figure 12** shows the changes in natural frequency and loss factor for a sandwich plate with an MRF core and MsM skins using Ramamoorthy et al. [22]. The results are evaluated for different skin thicknesses ht.(m). As seen in the figure, increasing the thickness of the MsM skins raises the natural frequency of vibrations and sharply reduces the loss factor. These results are based on applying a magnetic field of up to 5000 Gauss to the MRF core. When the core thickness remains constant and the skin thickness increases, the system's stiffness increases accordingly. This increase in stiffness naturally leads to an increase in the natural frequency.

In all the figures presented thus far, the effect of the magnetic field on the MsM skins was ignored to focus solely on the core behavior of the MRF. **Figure 13** provide a 3D view showing the simultaneous effect of the magnetic field on both the MRF core and the MSM skins.

Previously, it was observed that increasing the magnetic field in Ramamoorthy et al. [22] to 6000 Gauss raised the natural frequency of the circular sandwich plate. However, the current figure demonstrates that increasing the velocity feedback control parameter $K_{vfc}$ $(A.s/m^2)$ reduces the natural frequency, with different curves emerging under different magnetic field strengths. This reduction occurs because $K_{vfc}$ increases system damping, which lowers the natural frequency.

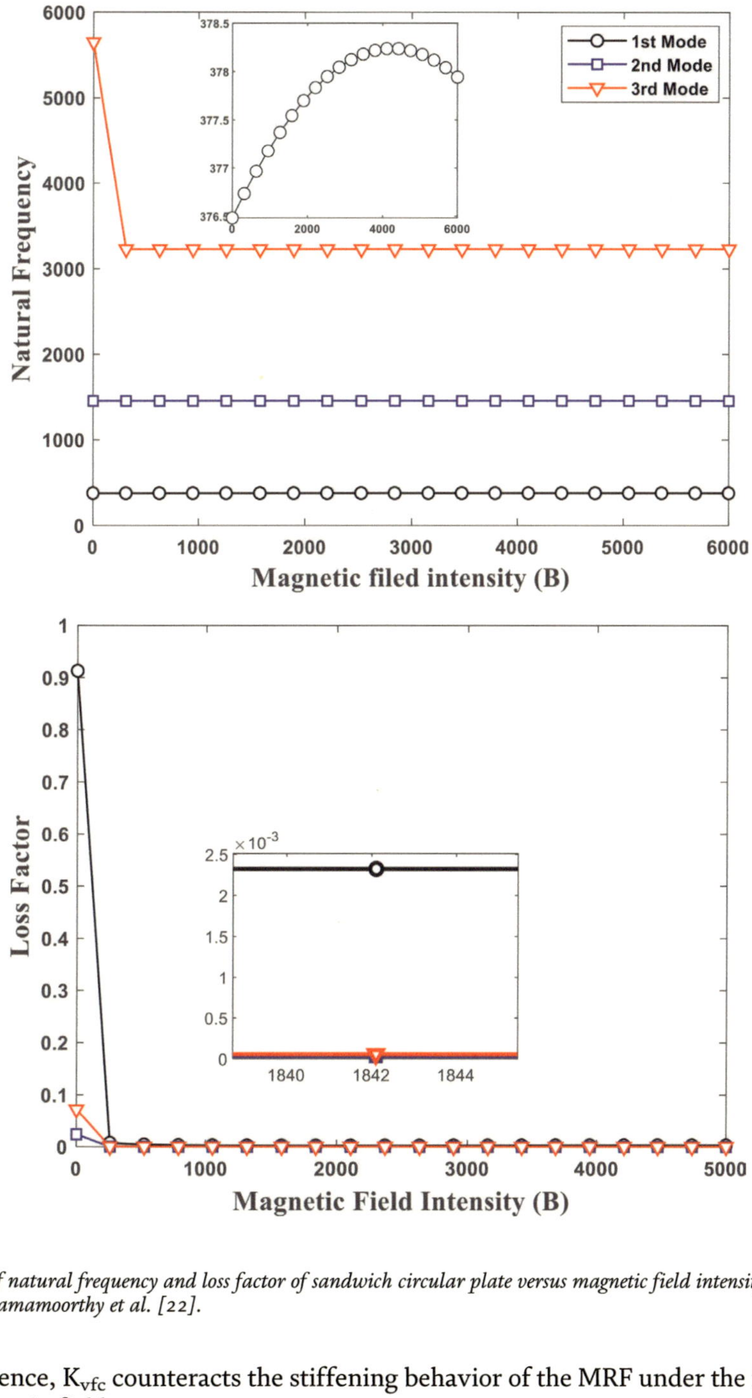


**Figure 9.**
*Variation of natural frequency and loss factor of sandwich circular plate versus magnetic field intensity in three modes for Ramamoorthy et al. [22].*

In essence, $K_{vfc}$ counteracts the stiffening behavior of the MRF under the influence of a magnetic field.

**Figure 14** illustrates the changes in natural frequency and the system's instability under the influence of non-conservative loads.

**Figure 14(a)** shows the overlap of two frequency modes as the non-conservative force increases, indicating the point where flutter occurs. Increasing the magnetic field in Ramamoorthy et al. [22] does not have a significant effect on system

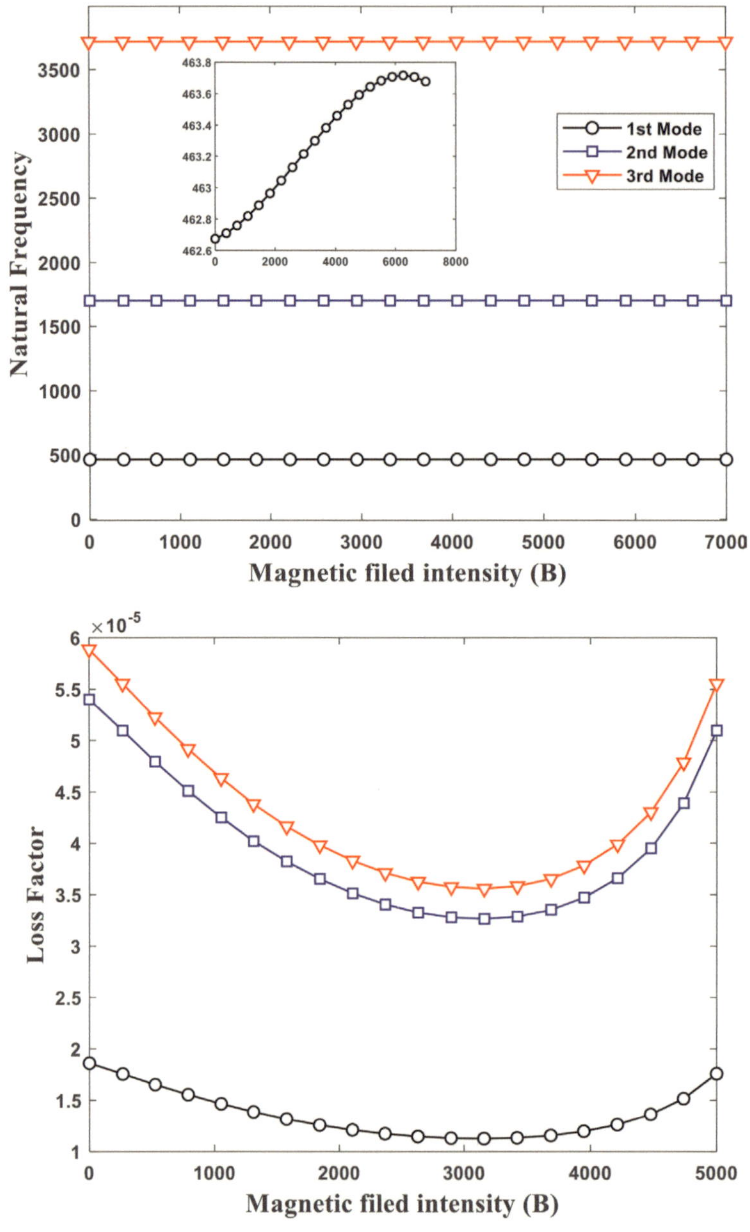


**Figure 10.**
*Variation of natural frequency and loss factor of sandwich circular plate versus magnetic field intensity in three modes for Aguib et al. [23].*

instability. However, this does not imply that the magnetic field is ineffective. Due to previously mentioned limitations, higher magnetic field values were not applied.

**Figure 14(b)** demonstrates the effect of non-conservative forces on the instability of the sandwich circular plate when the magnetic field in Ramamoorthy et al. [22] model is set to 100 Gauss. This figure also highlights the impact of $K_{vfc}$. As seen, increasing $K_{vfc}$ leads to a reduction in the natural frequency. This drop accelerates the onset of instability. However, similar to **Figure 14(a)**, these changes are not very

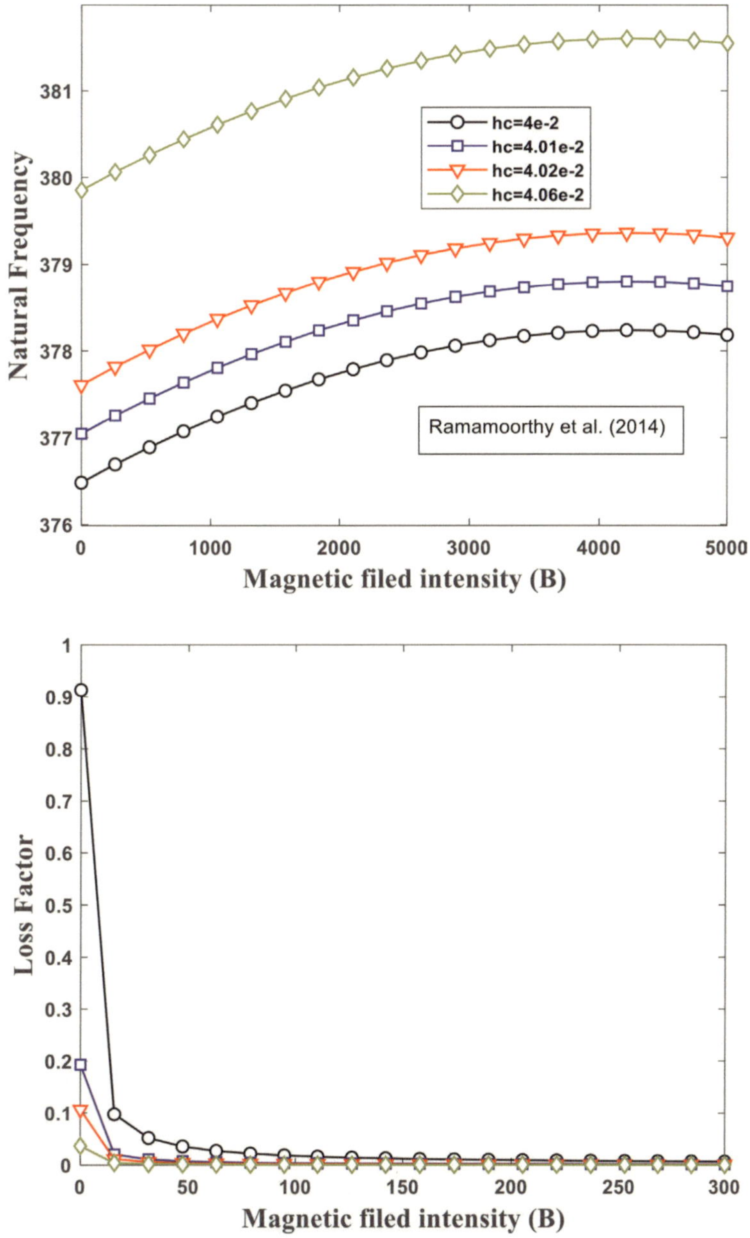


**Figure 11.**
*Variation of natural frequency and loss factor of sandwich circular plate versus magnetic field intensity in three modes for high-order model.*

pronounced. It is important to note that assigning much higher values to $K_{vfc}$ may yield different results.

### 5.3 Rectangular plate

In this section, the vibrations and frequency response of a sandwich square plate with an MRF core and MsM skins are investigated. The MRF core is modeled using

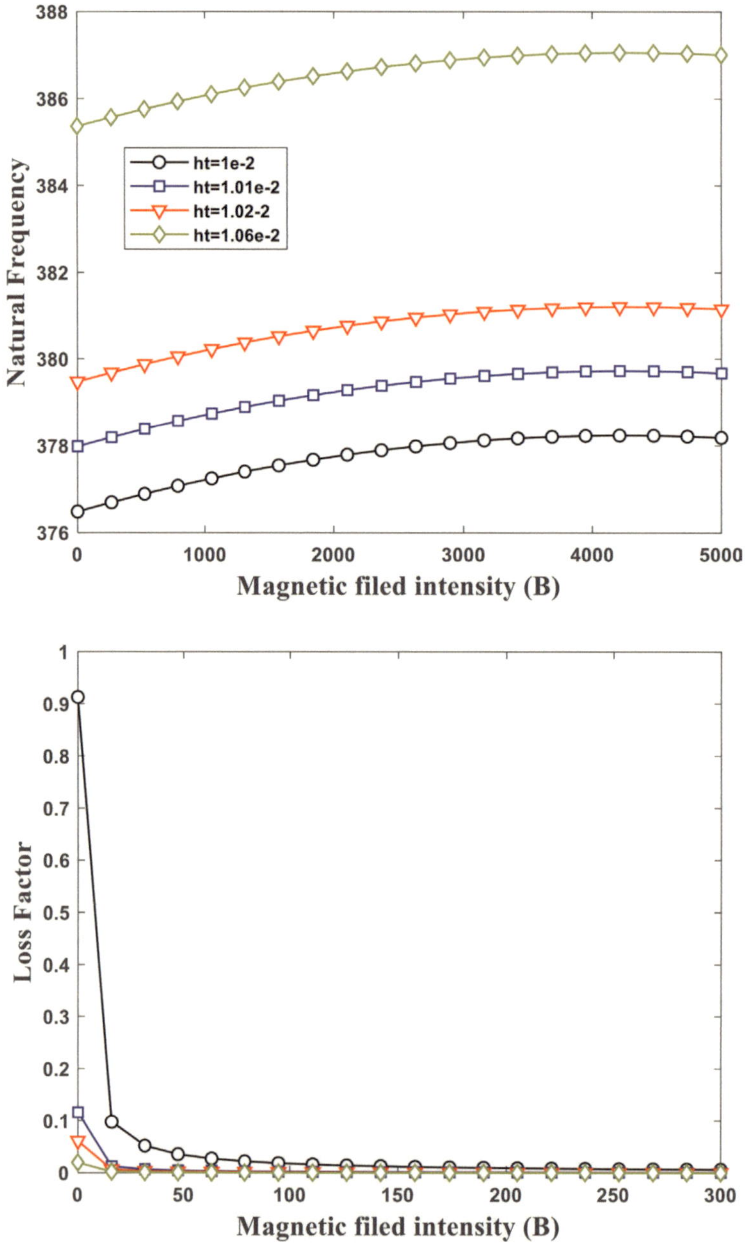


**Figure 12.**
*Variation of natural frequency and loss factor of sandwich circular plate versus magnetic field intensity in different thicknesses of skins.*

Ramamoorthy et al. [22] model, which did not yield reliable results for the square plate. The magnetic field values were selected based on the accurate responses from the presented models.

As shown in **Figure 15**, the magnetic field in Aguib et al. [23] model was increased to 1500 Gauss. The results indicate that, in the first three frequency modes, the natural frequency of vibrations increases with the magnetic field. This aligns well with

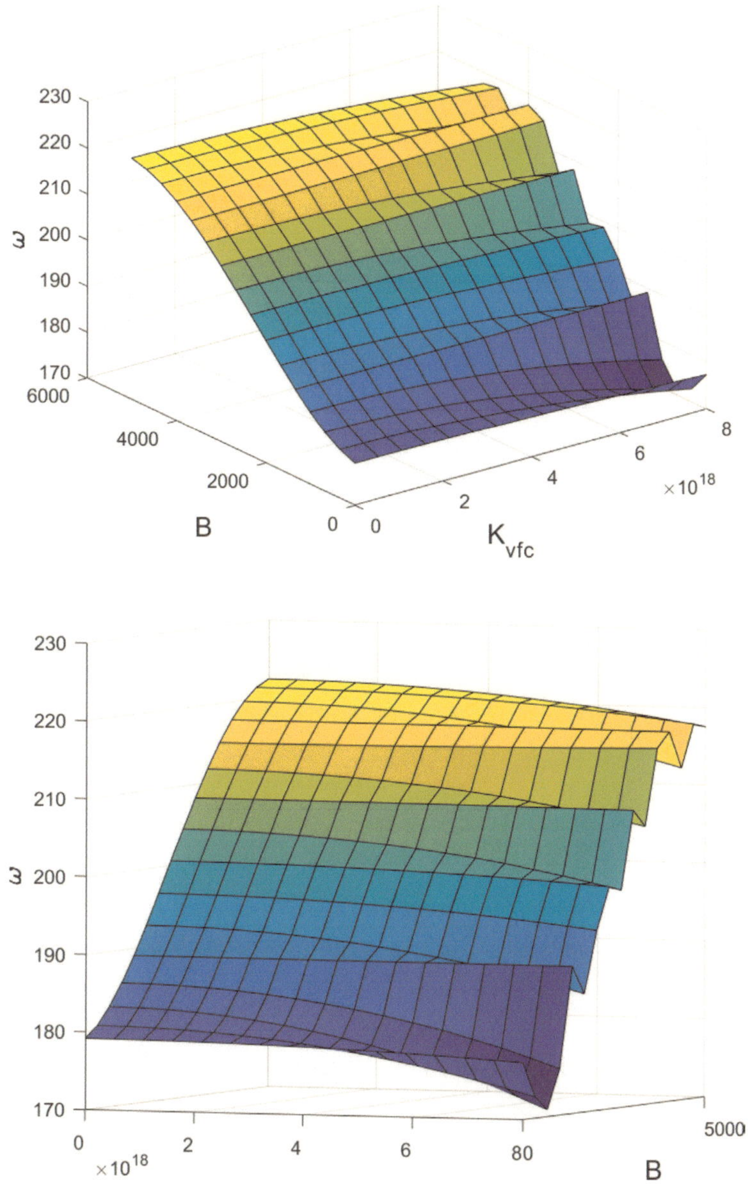


**Figure 13.**
*Variation of natural frequency sandwich circular plate versus magnetic field intensity and velocity feedback control parameter.*

earlier predictions since increasing the magnetic field stiffens the MRF core, enhancing the system's overall stiffness.

**Figure 15** (right) demonstrates that the loss factor in the third mode also increases. This result is consistent with expectations, as a review of other references shows that the loss factor can increase in certain cases.

**Figure 16** illustrates the changes in natural frequency with increasing magnetic field strength. **Figure 16** is plotted using the aspect ratio of the skin thickness to the core thickness. Two key observations can be drawn from this figure:

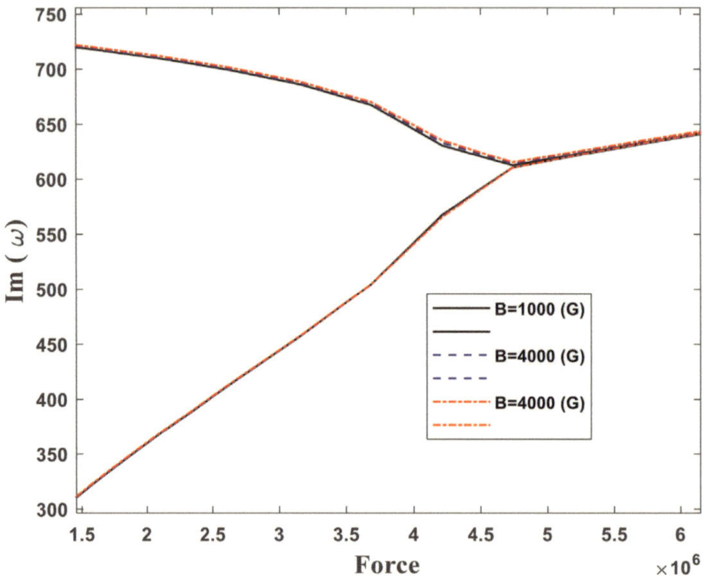


*(a)K_vfc=0*

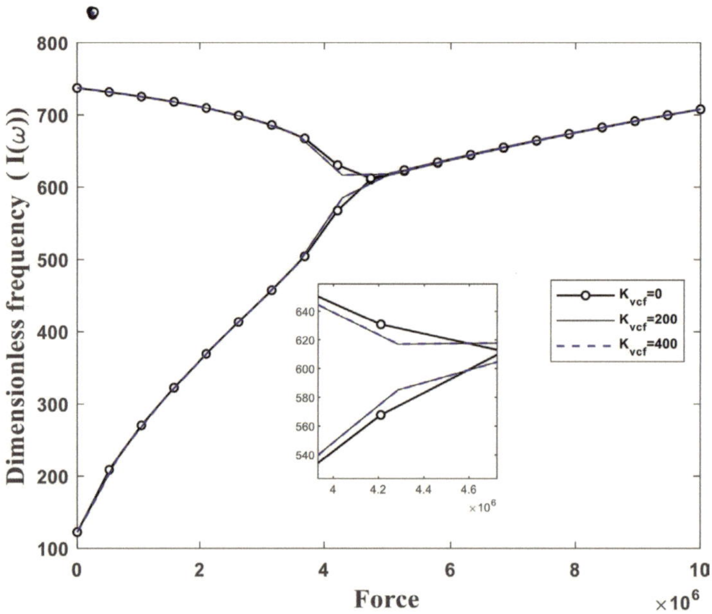


*(b)B=0*

**Figure 14.**
*Variation of real and imagery of frequency versus non-conservative force in different magnetic field intensity of sandwich circular plate for Ramamoorthy et al. [22]. (a) $K_{vfc}$ = 0. (b) B = 0.*

1. Increasing the aspect ratio—which implies increasing the thickness of the MsM skins—results in a higher natural frequency. This is because MsM skins have a high Young's modulus even without a magnetic field, and its addition to the sandwich structure increases the overall stiffness.

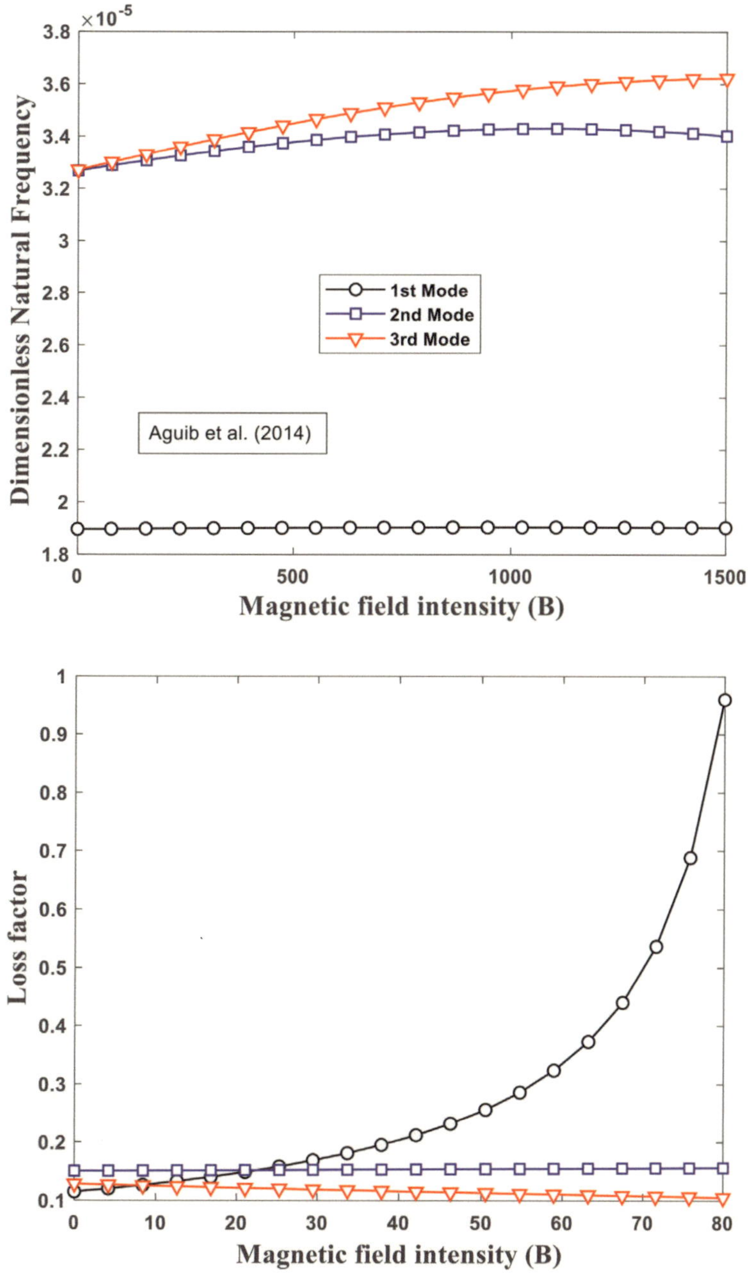


**Figure 15.**
*Variation of natural frequency and loss factor of sandwich rectangular plate versus magnetic field intensity in three modes for Aguib et al. [23].*

2. Effect of magnetic field on damping: If the applied magnetic field increases the $K_{vfc}$, the control system introduces additional damping, which can lead to a reduction in the natural frequency.

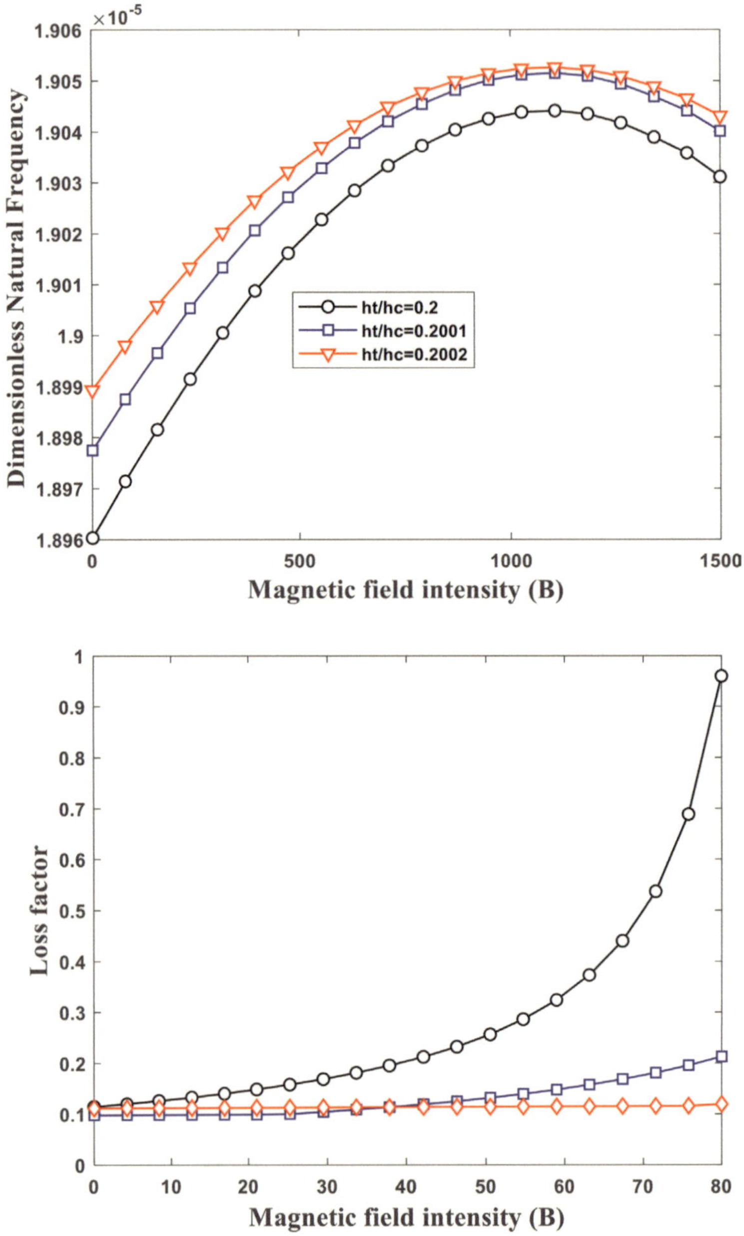


**Figure 16.**
*Variation of natural frequency and loss factor of sandwich rectangular plate versus magnetic field intensity for different aspect ratio.*

Thus, while increasing the skin thickness generally increases system stiffness and natural frequency, the influence of the magnetic field on $K_{vfc}$ and damping can counteract this effect.

**Figure 17** illustrates the combined effect of the applied magnetic field and the increase in $K_{vfc}$ on the MRF core and MsM skins. While the magnetic field stiffens the

*DOI: http://dx.doi.org/10.5772/intechopen.1009143*

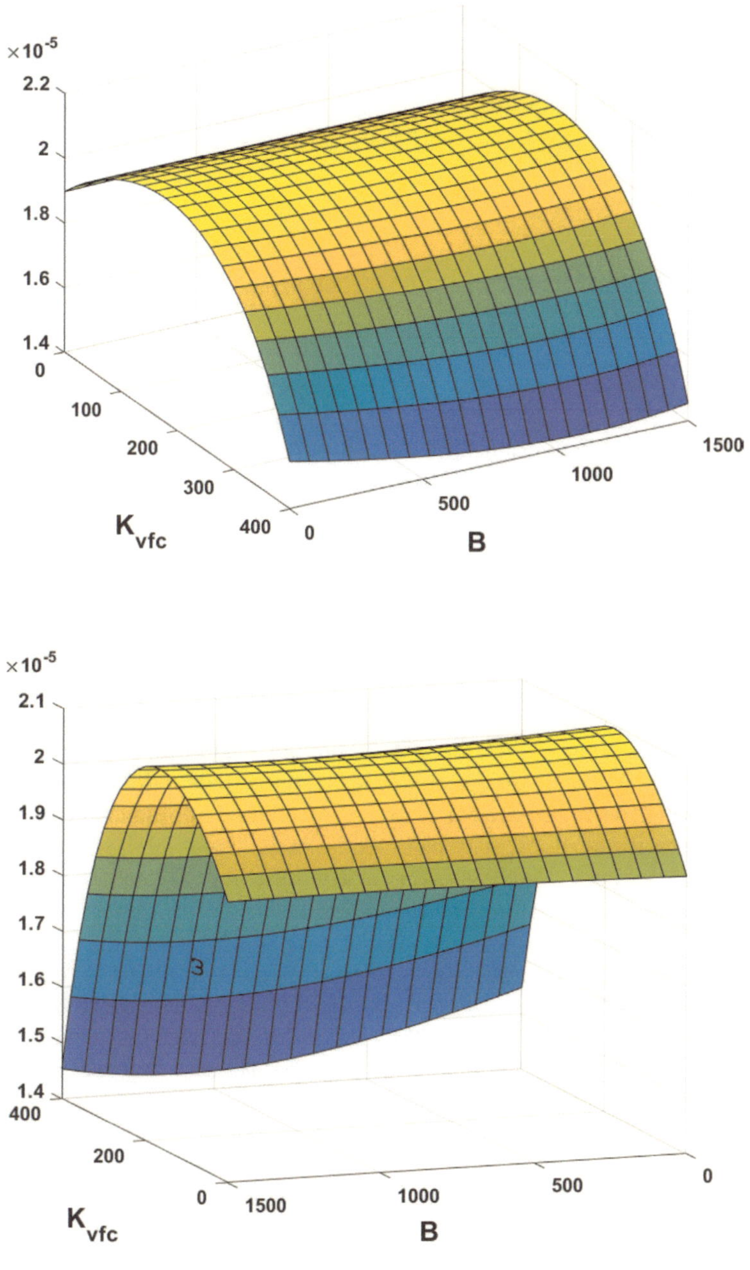


**Figure 17.**
*Variation of natural frequency of sandwich rectangular plate versus magnetic field intensity and velocity feedback control parameter.*

MRF core, making it behave more like a solid, the MsM skins exhibit increased damping when subjected to the same magnetic field.

As shown in the 3D view, increasing $K_{vfc}$ reduces the natural frequency due to enhanced damping in the MsM skins. In contrast, the magnetic field increases the stiffness of the MRF core, which tends to increase the natural frequency. The interaction between these two opposing effects is evident in the figures.

By analyzing the natural frequency values, it becomes clear that the velocity feedback control parameter has a more significant influence than the magnetic field applied to the fluid under the given conditions. This highlights an important consideration for designers: the interaction between MsM skins, and the MRF core subjected to the magnetic field, must be carefully accounted for to achieve optimal system performance.

## 6. Conclusion

This study presents a detailed vibration and instability analysis of three-layer sandwich structures, specifically focusing on beams, circular plates, and square plates with MsM surface layers and MRF cores. Three models of MRF (Ramamoorthy et al. [22], Ayub and Jacob) were introduced and tested under various magnetic field intensities and geometric configurations to identify the effect of magnetic fields on the natural frequency, loss factor, and system stability.

Key findings from the analysis include:

- *Effect of magnetic field on natural frequency and stiffness*: Across all tested structures, increasing the magnetic field led to a consistent rise in the natural frequency due to the stiffening of the MRF core. This effect was observed in Ramamoorthy et al. [22] model up to 6000 Gauss and in Aguib et al. [23] model up to 7000 Gauss.

- *Loss factor and damping behavior*: The loss factor (damping coefficient) typically decreases with increasing magnetic field strength. This reduction occurred because the stiffening of the MRF core reduced its damping capacity. However, for certain configurations, such as in Aguib et al. [23] model, the loss factor showed an unexpected increase when the magnetic field exceeded a threshold (e.g., 4000 Gauss).

- *Geometric parameter influence*: Variations in the core and skin thickness significantly influenced the natural frequency and damping behavior. Increasing the core thickness, while keeping the total sandwich thickness constant, increased the natural frequency due to higher stiffness, whereas the thickness of the MsM skins inversely affected this relationship.

- *Velocity feedback control parameter (Kvfc)*: The introduction of a velocity feedback control parameter (Kvfc) highlighted its opposing effect to the magnetic field. While magnetic fields increased stiffness and natural frequency, increasing $K_{vfc}$ enhanced damping and reduced the natural frequency. The interaction between these factors must be carefully managed to optimize system performance.

- *Instability and flutter*: The analysis of instability under non-conservative forces revealed that flutter occurs when vibration modes (such as the first and second modes) overlap. The magnetic field had a limited impact on preventing instability beyond certain intensities due to previously identified constraints.

Overall, the study offers valuable insights into the dynamic behavior of MRF-based sandwich structures, providing a foundation for future designs that demand enhanced vibration control and stability under varying loads.

The findings highlight the adaptability of smart-composite sandwich structures for engineering applications. Magnetic fields effectively tune the natural frequency and stiffness, which is useful for vibration isolation and adaptive systems. Damping behavior, influenced by magnetic field strength and geometry, offers design flexibility for energy dissipation or precision systems. The interplay between velocity feedback control (Kvfc) and magnetic fields enables precision tuning for dynamic stability and vibration suppression. Insights into flutter and instability provide guidance for safe and efficient designs in aerospace, marine, and automotive applications.

## Author details

Zahra Khoddami Maraghi
Faculty of Engineering, Mahallat Institute of Higher Education, Mahallat, Iran

*Address all correspondence to: z.khoddami@gmail.com

## References


[1] Nezami M, Gholami B. Optimal locations of magnetorheological fluid pockets embedded in an elastically supported honeycomb sandwich beams for supersonic flutter suppression. European Journal of Mechanics-A/Solids. 2019;**74**:81-95

[2] Ghorbanpour Arani A, Soleymani T. Size-dependent vibration analysis of an axially moving sandwich beam with MR core and axially FGM faces layers in yawed supersonic airflow. European Journal of Mechanics-A/Solids. 2019;**77**: 103792

[3] Khoddami Maraghi Z. Flutter and divergence instability of nanocomposite sandwich plate with magnetostrictive face sheets. Journal of Sound and Vibration. 2019;**457**:240-260

[4] Aboutalebi R, Eshaghi M, Taghvaeipour A. Nonlinear vibration analysis of circular/annular/sector sandwich panels incorporating magnetorheological fluid operating in the post-yield region. Journal of Intelligent Material Systems and Structures. 2021;**32**(7):781-796

[5] Sharif U, Sun B, Hussain S, Ibrahim DS, Adewale OO, Ashraf S, et al. Dynamic behavior of sandwich structures with magnetorheological elastomer: A review. Materials. 2021;**14** (22):7025

[6] Kuriakose VM, Sreehari VM. Experimental investigation on the enhancement of vibration and flutter characteristics of damaged composite plates using piezoelectric patches. Composite Structures. 2021;**275**:114518

[7] Amiri A, Talebitooti R, Aliakbari F. Size-dependent dynamics and instability of sandwich magnetorheological elastomer (MRE)-cored shells in presence of moving flow, based on modified first strain gradient theory. Thin-Walled Structures. 2022;**181**: 109800

[8] Sarafraz M, Seidi H, Kakavand F, Viliani NS. Flutter characteristics of a rectangular sandwich plate with laminated three-phase polymer/GNP/ fiber face sheets and an auxetic honeycomb core in yawed supersonic fluid flow. Journal of the Brazilian Society of Mechanical Sciences and Engineering. 2023;**45**(4):197

[9] Omidi Soroor A, Asgari M, Haddadpour H. On the dynamic behavior interpretation of sandwich beams with axially graded face sheets and magnetorheological core using modal strain energy approach. Acta Mechanica. 2023;**234**(7):2985-3008

[10] Oveissi S, Salehi M, Ghassemi A, Eftekhari SA, Ziaei-Rad S. Energy harvesting of nanofluid-conveying axially moving cylindrical composite nanoshells of integrated CNT and piezoelectric layers with magnetorheological elastomer core under external fluid vortex-induced vibration. Journal of Magnetism and Magnetic Materials. 2023;**572**: 170551

[11] Wan PH, Al-Furjan MS, Kolahchi R. Nonlinear flutter response and reliability of supersonic smart hybrid nanocomposite rupture trapezoidal plates subjected to yawed flow using DQHFEM. Aerospace Science and Technology. 2024;**145**:108862

[12] Moreira JA, Moleiro F, Araújo AL, Pagani A. Active aeroelastic flutter control of supersonic smart variable stiffness composite panels using

layerwise models. Composite Structures. 2024;**343**:118287

[13] Jonna N, Srinivas J. Aeroelastic stability analysis and optimization studies of rotating tapered composite sandwich blade with magnetorheological elastomer core. Journal of the Brazilian Society of Mechanical Sciences and Engineering. 2024;**46**(12):713

[14] Reddy JN. Theories and Analyses of Beams and Axisymmetric Circular Plates. Boca Raton, Florida, USA: CRC Press; 2022

[15] Mihankhah A, Khoddami Maraghi Z, Ghorbanpour Arani A, Niknejad S. Magneto-rheological response in vibration of intelligent sandwich plate with velocity feedback control. Journal of Solid Mechanics. 2022;**14**(4):364-377

[16] Eshaghi M. Supersonic flutter analysis of annular/circular sandwich panels containing magnetorheological fluid. Journal of Sandwich Structures & Materials. 2021;**23**(7):2968-2987

[17] Muhammad A, Yao XL, Deng ZC. Review of magnetorheological (MR) fluids and its applications in vibration control. Journal of Marine Science and Application. 2006;**5**:17-29

[18] Rezaeepazhand J, Pahlavan L. Transient response of sandwich beams with electrorheological core. Journal of Intelligent Material Systems and Structures. 2009;**20**(2):171-179

[19] Yeh JY. Vibration analyses of the annular plate with electrorheological fluid damping treatment. Finite Elements in Analysis and Design. 2007;**43**(11–12):965-974

[20] Khoddami Maraghi Z, Amir S, Arshid E. On the natural frequencies of smart circular plates with magnetorheological fluid core embedded between magnetostrictive patches on Kerr elastic substance. Mechanics Based Design of Structures and Machines. 2024;**52**(3):1651-1668

[21] Lee CY, Cheng CC. Complex moduli of electrorheological material under oscillatory shear. International Journal of Mechanical Sciences. 2000;**42**(3):561-573

[22] Ramamoorthy M, Rajamohan V, Jeevanantham AK. Vibration analysis of a partially treated laminated composite magnetorheological fluid sandwich plate. Journal of Vibration and Control. 2016;**22**(3):869-895

[23] Aguib S, Nour A, Zahloul H, Bossis G, Chevalier Y, Lançon P. Dynamic behavior analysis of a magnetorheological elastomer sandwich plate. International Journal of Mechanical Sciences. 2014;**87**:118-136

[24] Chen L, Gong X-L, Jiang W-Q, Yao J-J, Deng H-X, Li W-H. Investigation on magnetorheological elastomers based on natural rubber. Journal of Materials Science. 2007;**42**(14):5483-5489

[25] Kumar JS, Ganesan N, Swarnamani S, Padmanabhan C. Active control of beam with magnetostrictive layer. Computers & Structures. 2003;**81**(13):1375-1382

[26] Hong CC. Transient responses of magnetostrictive plates without shear effects. International Journal of Engineering Science. 2009;**47**(3):355-362

[27] Hong CC. Transient responses of magnetostrictive plates by using the GDQ method. European Journal of Mechanics-A/Solids. 2010;**29**(6):1015-1021

[28] Murty AK, Anjanappa M, Wu YF. The use of magnetostrictive particle

actuators for vibration attenuation of flexible beams. Journal of Sound and Vibration. 1997;**206**(2):133-149

[29] Amir S, Arshid E, Khoddami Maraghi Z, Loghman A, Ghorbanpour AA. Vibration analysis of magnetorheological fluid circular sandwich plates with magnetostrictive facesheets exposed to monotonic magnetic field located on visco-Pasternak substrate. Journal of Vibration and Control. 2020;**26**(17–18):1523-1537

[30] Ghorbanpour Arani A, Khani Arani H, Khoddami Maraghi Z. Vibration analysis of rectangular magnetostrictive plate considering thickness variation in two directions. International Journal of Applied Mechanics. 2015;7(04):1550059

[31] Reddy JN. Energy Principles and Variational Methods in Applied Mechanics. Hoboken, New Jersey, USA: John Wiley & Sons; 2017

[32] Wu JS. Analytical and Numerical Methods for Vibration Analyses. Hoboken, New Jersey, USA: John Wiley & Sons; 2013

[33] Shu C. Differential Quadrature and its Application in Engineering. Dordrecht, Netherlands: Springer Science & Business Media; 2012

[34] Nayak B, Dwivedy SK, Murthy KS. Dynamic stability of magnetorheological elastomer based adaptive sandwich beam with conductive skins using FEM and the harmonic balance method. International Journal of Mechanical Sciences. 2013;77:205-216

[35] Asgari M, Kouchakzadeh MA. Aeroelastic characteristics of magneto-rheological fluid sandwich beams in supersonic airflow. Composite Structures. 2016;**143**:93-102

Chapter 3

# Vibration in Metal Cutting Process

*Cătălin Gabriel Dumitraş, Mihaiţă Horodincă, Adriana Munteanu, Florin Chifan, Dragoş-Florin Chitariu and George-Gabriel Chiriac*

## Abstract

The cutting processes occupy a special place within the manufacturing processes, 70% from the total manufacturing processes are represented by cutting processes. Moreover, within the cutting processes, turning and milling processes are mostly used, therefore we will focus on this type of processes. The cutting process is complex due to the multitude of factors such as the geometry of the tool, the discontinuous cutting regime, the material of the active cutting part and the workpiece, the temperatures in the cutting area, the instability of the system machine tool-devices-cutting tool-part. These parameters lead to a complex dynamics of the process with major influences in the optimization of criteria such as productivity, the quality of the processed surfaces, consumption of cutting tools (due in particular to the wear of the cutting tool). In this chapter, we emphasize the influence of the cutting parameters on the vibrations in the system, which influence the previously mentioned criteria. One present how to reduce the level of vibrations in the system either by changing the cutting regimes or by using active/passive systems.

**Keywords:** vibration, cutting tool geometry, tool life, wear, cutting force, surface quality, hardness, chatter

## 1. Introduction

### 1.1 Typology of vibrations in the cutting process

Machining by cutting is a complex surface generation process, in which the technological system machine-tool-device-tool-part (MDTP) is a predominantly elastic system. The entire system is subject to the action of exciting forces, with different intensities and frequencies.

The dynamic phenomena in machine tools are determined by the interaction between the elastic system and the manufacturing processes, which accompany the operation of the respective machine tools.

The elastic system consists of the actual machine tool, the clamping device, the cutting tool and the workpiece.



IntechOpen

The work processes represent the set of physical phenomena, which take place in the contact areas of the elements of the elastic system [1, 2]. Thus, the cutting process takes place in the contact area between the part and the tool; the friction process takes place between the elements of kinematic couples and joints; the processes in drive motors take place in the space between the rotor and the stator, between the piston and the cylinder, etc.

The elastic system and the manufacturing processes, in their interaction, are treated as a dynamic system specific to the machine tool.

The actions of manufacturing processes on the elastic system are generally of a mechanical nature (forces or moments), but they can also be of a thermal, electrical, etc.

As a result of these actions, periodical relative displacements of the structural elements of the elastic system occur, such as between the cutting tool and the workpiece; slide and guide; spindle and bearing; rotor and stator; etc. These displacements represent the reaction of the elastic system to the action of the cutting processes and lead to the variation of work parameters such as the chip section, the pressure on the contact surfaces, the relative displacement speed and are called vibrations.

In the process of mechanical machining by cutting, the occurrence of vibrations is inevitable, and in situations where their amplitude exceeds the limits imposed on the dimensional and shape precision of the generated surfaces, the vibrating phenomenon is harmful.

The main types of vibrations that occur in the process of mechanical machining by cutting are:

*Free vibrations.* These vibrations are caused by a short-term excitation. Free (transient) vibrations characterize the structural damping and regularly have a very short duration due to the high damping in the machine tool joints, which is why in most scientific articles [2–4] they are considered insignificant. Therefore, transient vibrations are those vibrations that appear temporarily in the process of transition to the stable operating state. Transient processes that are of particular interest in machining on machine tools are: the entry and exit of the tool into/from the cutting, the acceleration and braking of moving elements, the reversal of the direction of movement, etc.

*Forced vibrations.* These vibrations are caused by a periodic excitation phenomenon and have the same fundamental frequency as the excitation [2, 5]. In turn, forced vibrations can be:

- Forced vibrations that do not depend on the cutting process. Their occurrence is related to the action of inertial forces, which usually arise as a result of the rotational movement of unbalanced masses in the kinematic chains of machine tools, as well as the vibrations transmitted from other machines and neighboring installations [2]. This category of vibrations appears as a result of technological imperfections in the processing and assembly of machine tools, improper fixing of machine tools on the foundation. Forced vibrations that do not depend on the cutting process can also be caused by errors in the gears, unevenness of the belts or undulations in the bearing raceways.

- Forced vibrations that depend on the cutting process. Their occurrence is related to numerous factors such as: variation in the depth of cut; variation in the chip section, in the case of milling and broaching; periodic variation in the chip section when turning eccentric workpieces and when machining with eccentric grinding wheels; variation in the cutting speed [2, 5].

The dependence of this category of vibrations on the particularities of the technological process makes it possible to avoid their occurrence only in very rare cases (by reducing the variation in the machining depth of cut of the workpiece and the eccentricity of the tools).

*Self-vibrations*. Self-vibrations occur exclusively during the cutting process, due to the variation of the cutting force, the variation of the chip thickness and last but not least due to the inertia of the cutting process [2, 5]. Self-vibrations are sustained vibrations due to some exciting factors, generated by the vibratory movement itself (also called self-excited vibrations). According to the nature of the exciting factors, self-vibrations that occur in the elastic systems of machine tools can be:

- Self-vibrations that occur during the cutting process as a result of the interdependence between the magnitude of the cutting force and the relative displacement between the tools and the workpiece;

- Self-vibrations due to friction processes, through the nature of the dependence between the friction force and the sliding speed in the kinematic torques;

- Self-vibrations due to the non-uniformity of the machining allowance, which produce the variation of the cutting force and the non-uniform displacement between the tools and the workpiece.

Due to vibrations in the cutting process, there is wear on the machine tool, premature wear of the cutting tool, especially when using tools with metal carbides or mineral-ceramic inserts, and perhaps most importantly a worsening of the roughness of the manufactured surface, which is unfavorable for finishing operations. Along with the previous shortcomings, vibrations avoidance measures reduce the possibilities of full exploitation of the machine tool and produce noise pollution of the environment.

Numerous studies and researches that have been carried out in relation to the influence of vibrations on the cutting tool have led to the conclusion that in conditions of vibration, the tool durability, which represents the cutting duration between two re-sharpening or changing of the active cutting edge, can decrease by four times or even more [2, 4, 5].

One of the most known but undesirable phenomenon in cutting process, especially in boring and milling, is chatter.

Chatter occurs when a mechanical system (e.g., a machine tool) experiences self-sustaining vibrations due to the interaction between the cutting tool and the material being machined. Essentially, these vibrations become self-amplified, making them unstable and persisting for a longer period of time.

These vibrations can be considered self-vibration in the sense that they are generated by the system's own dynamic properties (such as the natural frequencies of the machine tool and the resonance that occurs during the process).

The difference is that self-vibration can be a general phenomenon, in which any type of self-vibration of a system is involved, without necessarily being related to an external process (such as the machining of the material). In contrast, chatter occurs specifically in the context of machining or cutting processes, being the result of an unstable interaction between the cutting tool and the material.

In other words, chatter is a type of self-vibration, but on the condition that those vibrations are induced by interaction with the processing process and become unstable, self-sustaining.

## 1.2 The main causes of vibrations in the cutting process

The appearance of vibrations in the cutting process is the consequence of the action of a multitude of exciting factors.

The most important causes of vibrations in the cutting process are summarized in **Figure 1** [1, 6].

### *1.2.1 Causes external to the system – Machine-device-tool-part (MDTP)*

The causes external to the elastic system are represented by the vibrations transmitted through the foundation of the machine tool. These can come from other neighboring machine tools, hammers, presses, compressors, unbalanced fans, etc. As a result of the vibrations transmitted through the ground from other machines, the vibrations of the floor of a workshop include a wide spectrum of frequencies [7]. Due to this fact, in general one of these vibrations can have a frequency close to the natural frequency of the machine tool on which the chip is being cut, thus forced vibrations will appear on it.

These vibrations generally have a small amplitude and are annoying, especially in the case of finishing machining, (precision boring, grinding) producing high roughness [6, 7].

### *1.2.2 Internal causes of the system – Machine-device-tool-part (MDTP)*

Inside the elastic system – machine-device-tool-part (MDTP) discontinuities or non-uniformities of movement occur, which lead to vibrations of the system [7].

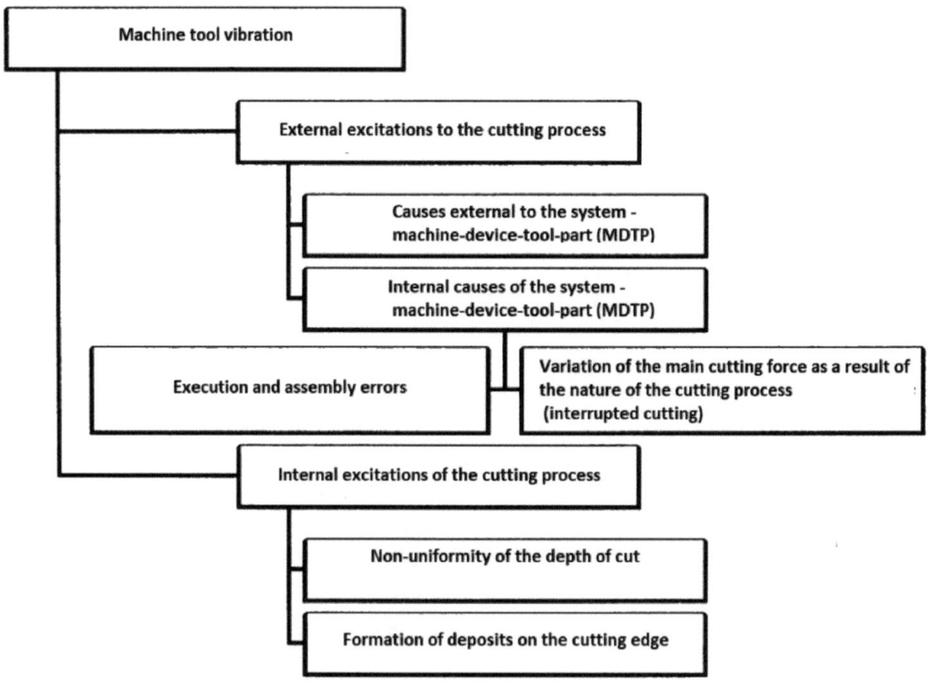


**Figure 1.**
*The most important causes of vibrations in the cutting process.*

The causes of these vibrations can be due to:

- *Non-uniformity of the material to be processed*. If the material to be processed has small hard grains, they can produce free vibrations on the semi-finished product and the tool. If these vibrations are quickly damped, their production is not important. If these transient vibrations are not damped, then large amplitude vibrations can be produced (due to dynamic instability), which become inconvenient [6, 7]. Non-uniformities in the cutting process can also be produced due to deposits on the cutting edge [1], deposits that periodically break off, causing the appearance of micro-irregularities (craters), which are relatively large [7].

- *Intermittent cutting*. When machining discontinuous surfaces (machining shafts with keyways, tool entry and exit into/out of the cutting - in the case of milling cutters, etc.) variations in the main cutting force occur, which cause vibrations. If these vibrations have a frequency that coincides with the natural frequency of the structure of the machine-device-tool-part system (MDTP), due to the resonance phenomenon, forced vibrations of high amplitude may appear [6, 7]. When machining with tools equipped with a single cutting tooth, these vibrations, although they influence the durability of the tool, do not have a high amplitude and are quickly damped. The same does not happen in grinding, where after the abrasive disc passes over a channel, the vibrations are not damped and marks appear on the workpiece. Also, the variation of the milling forces produces a variation of the torque necessary for milling, which gives rise to forced torsional vibrations on the milling cutter shaft. These vibrations have a negative influence on the durability of the milling cutter (especially for milling cutters reinforced with metal carbide inserts) [6, 7].

- *Unbalanced masses in rotational motion*. Due to the imprecision of execution or assembly of the machine-device-tool-part system (MDTP), which could contain: eccentricities of some machine parts, eccentricities of the semi-finished product, cutting tool runouts, etc., if their speed approaches one of the natural frequencies of the system structure, high-amplitude vibrations appear (due to the resonance), which negatively influence both the quality of processing and the durability of the tool [6].

- *Gear wheel deviations.* Gear wheel deviations, which could be: eccentricities, tooth eccentricity, division errors, tooth profile errors, etc., are the causes that can produce vibrations of the machine-device-tool-part system (MDTP). For this reason, in the case of precision processing (for example, grinding), gear wheels are replaced by transmission belts, which play the role of filters [6].

- *Belt drive.* Due to the variation of the effective radius of the belt (neutral fiber radius), the variation of the effective radius of the belt wheel, as well as due to structural defects of the belt (variation of the belt profile, inhomogeneity of the belt material, stiffness, etc.), forced torsional or rectilinear vibrations can occur [6, 7]. To attenuate these vibrations, trapezoidal belts are usually replaced by flat belts.

- *Bearings.* Due to the imprecision of the components of ball or roller bearings (rolling elements or raceways), vibrations can occur in the operation of the bearings, respectively vibrations during machining, which are undesirable in the case of high-precision cutting [6]. The best results are obtained when two rows of bearings are provided, with reduced radial clearance or even with negative radial clearance. In the latter case, an improvement of the machined surface is obtained, but the bearings heat up and their durability decreases. Vibrations also occur in the case of sliding bearings, when there are dimensional inaccuracies or roundness due to wear. Sliding bearings have a greater damping capacity than that of roller bearings; this is true when a lubricant with high viscosity is used. With the increase in the temperature of the lubricant, the damping capacity decreases [8].

- *Nonlinearity of the cutting force in relation to the cutting speed.* The nonlinearity of the cutting force in relation to the speed is one of the causes of the appearance of self-vibrations in machining. The main component *Fc* of the cutting force is determined by the relation [5]:

$$F_c = C_F \cdot K_F \cdot a^{XF} \cdot s^{yF} \cdot v^{-nF} \tag{1}$$

where:

$C_F$—constant of the material subjected to machining, having a certain geometry determined under certain cutting conditions;

$K_F$—correction coefficient depending on the specific machining conditions;

$a$—cutting depth [mm];

$s$—feed rate [mm/rev];

$v$—cutting speed [m/min];

$x_F$, $y_F$, $n_F$—influence exponents of the cutting regime parameters.

The variation of the cutting force in relation to the cutting speed is nonlinear and decreasing, as can be seen in **Figure 2** [9].

- *Cutting process inertia.* The machining process inertia introduce a time delay in the variation of the cutting force, in relation to the relative displacement between the tool and the workpiece, reported to the cutting tool tip. In other words there is a phase shift between the variation of the force and the relative position of the tool-workpiece (and also its derivative as velocity). The value of this phase shift can generate a positive feedback force-velocity which is the main reason for cutting instability (or appearance of self-excited vibrations) [1].

### 1.3 Factors that cause vibrations in the cutting process

If vibrations are detected during the cutting process, the first operation that is required is to disconnect the machine tool. If after disconnecting the machine tool the vibrations cease, then their cause is inside the elastic system machine-device-tool-part (MDTP), and if the vibrations persist, then their cause is due to vibrations transmitted from other neighboring machines, through the foundation of the machine tool on which the cutting is being done. Then the source of the vibrations is traced, comparing the frequency and sometimes the amplitude of the vibrations from the machine tool, with the frequency and amplitude measured at different disturbing places. If it has been established that the vibrations are due to the machine-device-tool-part system (MDTP), then the machine tool is coupled [2]. If the vibrations cease, then their cause is due to the cutting process. Otherwise, if strong vibrations appear on the

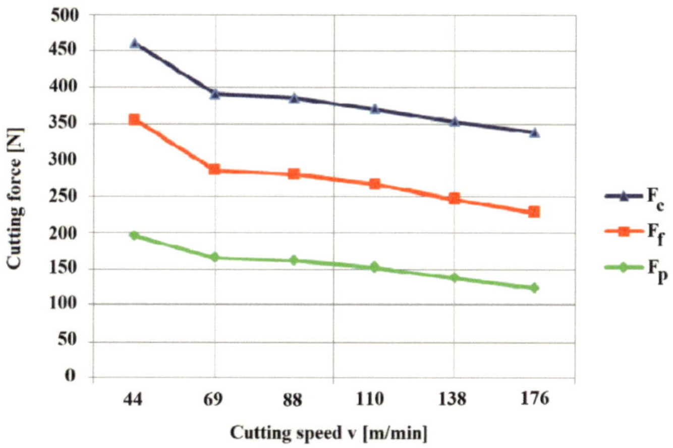


**Figure 2.**
*Variation of the cutting force in relation to the cutting speed (adapted from [9]).*

machine tool, then they come from the various unbalanced masses in motion, following the frequency of the vibrations. The frequency of these vibrations is compared with the frequency of the vibrations of the various parts of the machine tool.

If it is proven that the vibrations come from the cutting process, then their frequency and the frequency of the tool's entry into cutting are established. If, when the speed changes, the vibration frequency remains equal to that of the cutting interruptions, the cause of the vibrations. However, if, when the speed changes, the vibration frequency differs from that of the cutting interruptions and does not change or changes insignificantly when the speed varies, then the vibrations are self-excited.

The main factors influencing the production of vibrations are:

- *Workpiece material.* Workpiece materials influence vibrations through their damping capacity [6, 10]. Of all the materials, gray cast iron has the highest vibration damping capacity, followed by nodular cast iron and then cast steel. Also, the toughness of the workpiece material has an influence on the occurrence of vibrations. The strongest vibrations occur when machining tough steel, because in this case the relationship between friction force and cutting speed is quasi-proportional. Heat treatment and the structure of the workpiece material also influence the damping properties of the materials. For example, as the hardness of gray cast iron increases, its damping capacity may decrease; cast irons with a pearlitic structure have a lower damping capacity than cast irons with a ferritic structure. It has also been experimentally found that with increasing material hardness, the vibration amplitudes decrease almost linearly. Thus, **Figure 3** shows the influence of the hardness of the material to be processed on the amplitude and frequency of vibrations that occur during cutting, for different materials, from non-ferrous (Al, Bronze) to cast iron and steel.

- The explanation is that, by improving the structure of the material to be processed (which means a reduction of the grains by heat treatment for improvement), the plastic deformation in the cutting zone is lower. It can be seen that the group of steels presents values for the amplitude and frequency of vibrations lower than the non-ferrous ones. In **Figure 4** the influence of the tension of the

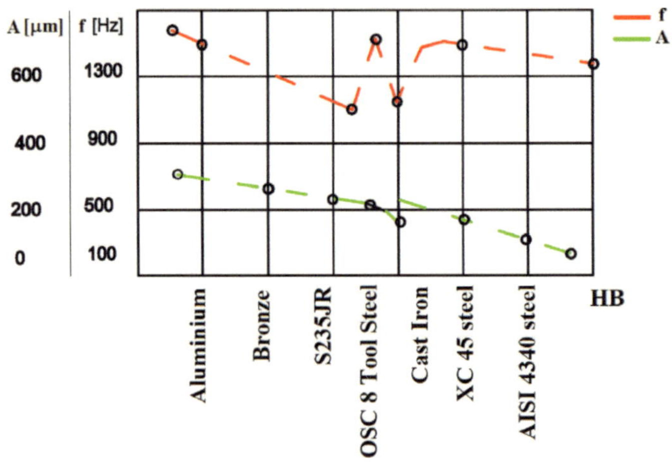


**Figure 3.**
*The influence of the hardness of the material (adapted from [10]).*

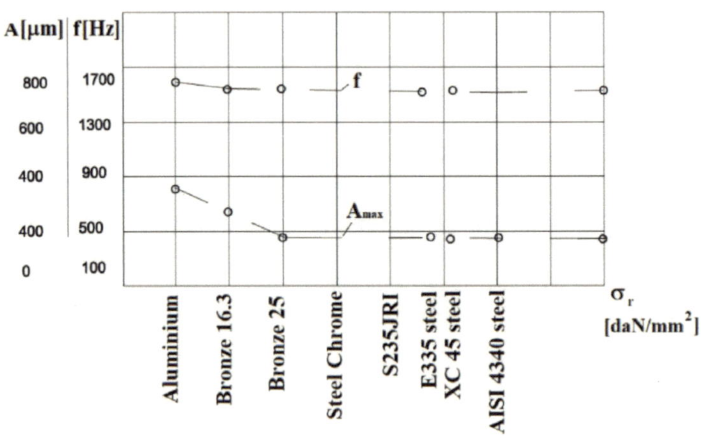


**Figure 4.**
*The influence of the tension of the material (adapted from [6]).*

material to be processed on the amplitude, respectively the frequency of vibrations that occur during cutting is presented [6]. It is observed that the increase in carbon content and the content of alloying elements contributes to a significant reduction in the amplitude of vibrations. Also, it is found that with the increase in elongation, the amplitude of vibrations increases sharply, and their frequency increases quasi-linearly. This shows that non-ferrous metals have an elongation favorable to high amplitudes and frequencies.

For a given material, for example XC 45, it is observed that with increasing cutting speed, the friction coefficient between the tool rake face and the chip decreases (**Figure 5**), due to the fact that the adhesion force of the chip on the clearance face is low and in addition the temperature in the cutting zone is high [6].

- *Tool material and tool surface precision.* The tool material and tool surface precision influence vibrations, to the extent that they have repercussions on the

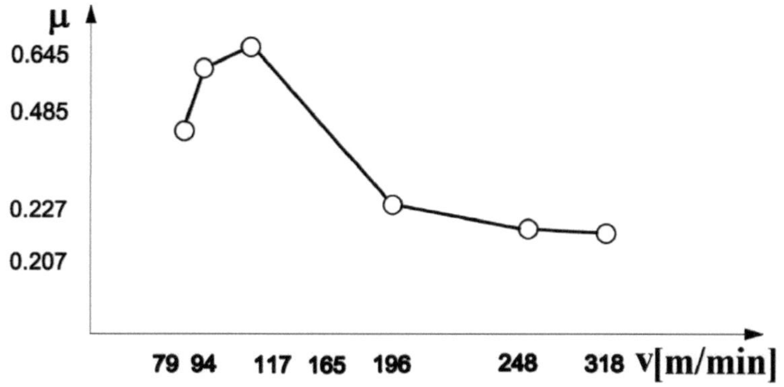


**Figure 5.**
*The influence of the cutting speed on friction coefficient.*

coefficient of friction between the chip and the tool clearance face, that is, to the extent that they influence the values of the friction forces [10]. With the increase in friction forces, vibrations are amplified.

- *Tool geometric parameters and tool position.* The vibrations that occur in the cutting process are also greatly influenced by the geometric parameters, through the clearance angle α, side rake angle γ, the nose radius $r_e$, and the position parameters, through the side cutting edge angle $K_r$, [6]. Theoretically, the shape of the tool must be chosen so as to increase the cutting resistance of the material to be processed. **Figure 6** shows the influence of the side rake angle γ on the amplitude of vibrations that occur during machining.

Analyzing the influence of the nose radius on the amplitude, respectively the frequency of vibrations, it is observed that the frequency does not depend on the nose

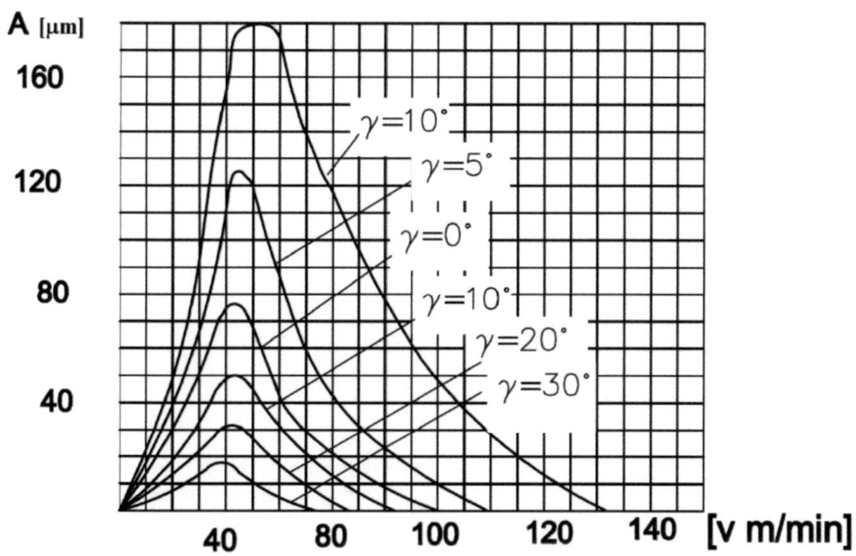


**Figure 6.**
*The influence of the side rake angle γ on the amplitude of vibration (adapted from [6]).*

radius of the tool, and the amplitude increases with the nose radius, more sharply for relatively small radius up to 6 mm. For nose radius of the cutting tool greater than 6 mm, the increase in amplitude is smoother. It is observed that the frequency is not influenced by the side cutting edge angle, and the amplitude of the vibrations decreases with increasing side cutting edge angle. **Figure 7** shows the wear characteristic of a turning tool, the change in the amplitude and frequency of vibrations, depending on the cutting time.

The wear characteristic has three zones: initial wear, normal wear and increased wear (final). Regarding the amplitude of the vibrations, a first zone of relatively smooth growth is observed, after which from a certain critical cutting time the amplitude increases exponentially, and after exceeding the threshold, a smooth growth is observed again.

- *Cutting regime parameters.* The occurrence of vibrations in the cutting process is also greatly influenced by the cutting regime, through the three parameters: cutting speed v, feed rate s and cutting depth a. In machining operations with cutting, high feed rates and high cutting depths favor the occurrence of vibrations, due to the large cross-section of the chips and due to the strong stress on the moving elements of the machine tools (feed box, main spindle support bearings). In most cases, vibrations are influenced by the cutting speed. From the analysis of the dependence between the amplitude of vibrations and the cutting speed it is found that there is a critical zone, in which the occurrence of vibrations is favored.

- *Coolant-lubricant.* The coolant-lubricant influences the occurrence of vibrations in the cutting process, to the extent that it influences the coefficient of friction between the chip and the tool clearance face [6]. When the coefficient of friction decreases, respectively the friction forces decrease, the vibrations are reduced.

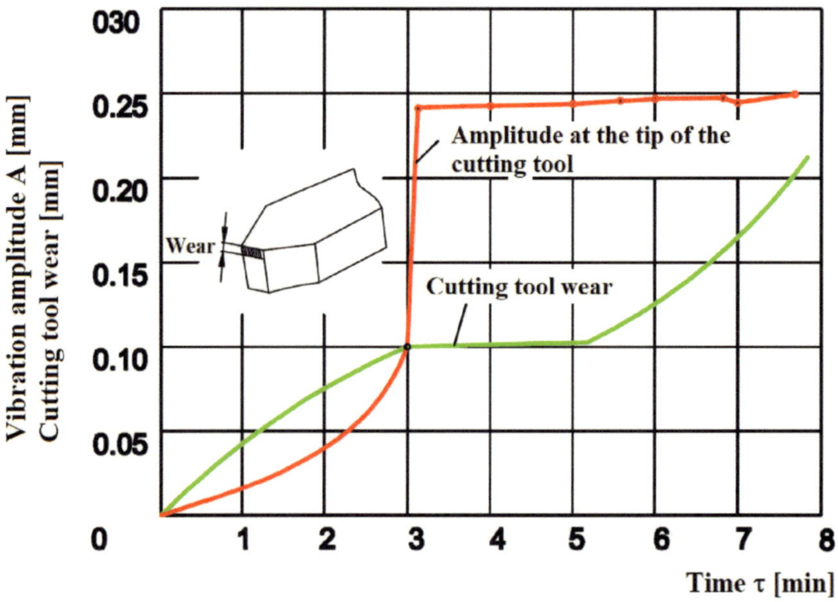


**Figure 7.**
*The wear characteristic of a turning tool.*

In the mechanical process before cutting, the occurrence of vibrations is inevitable. These vibrations also cause vibrations in the machine tool on which the chip is being cut, premature wear of the cutting tool, worsening of the roughness of the machined surface, reduction of the integral exploitation of the machine power as well as noise pollution.

In situations where the amplitude of these vibrations exceeds the limits imposed on the dimensional and shape precision of the generated surfaces, the vibration phenomenon becomes harmful and therefore they must be attenuated.

## 2. Chatter in metal cutting processes

As we mention above the self-excited vibration, called chatter, is one of the main limitations in metal removal processes that is why we will pay more attention to this phenomenon.. Chatter may spoil the surface of the part and can also cause large reduction in the life of the different components of the machine tool including cutting tool itself. This undesirable vibration during machining operations is unpredictable and the consequences are very costly.

The occurrence of chatter during machining can have several detrimental effects:

- *Reduced surface quality:* Vibrations caused by chatter can lead to an irregular machined surface, with visible ripples or scratches, affecting the desired finish.

- *Rapid tool wear:* Continuous vibrations can cause faster wear of cutting tools, leading to the need for frequent tool changes and additional costs.

- *Process instability*: Chatter can make the machining process unstable, which can lead to fluctuations in the depth of cut or variations in the feed rate. This can reduce the efficiency and accuracy of the operation.

- *Equipment damage:* Intense vibrations can lead to additional stress on machine tool components, which can cause damage to the equipment or affect its long-term durability.

- *Reduced productivity:* Although chatter may seem like a minor phenomenon at first, in reality, it can significantly reduce machining efficiency, requiring additional adjustments, downtime for repair, or even complete process shutdown.

Every spindle, toolholder and tool has a set of frequencies at which it naturally wants to vibrate. Each tool has a specific set of feeds and cutting speeds that will match one of its natural frequencies and the result will be a smooth cut with minimal chatter. Finding the right set of parameters for each tool assembly may take some experimentation but the result will be the ability to maximize the performance of the machine tool.

If one experience sudden chatter the first thing that must be done is to increase the rigidity of the tool assembly, this can be obtained by using the shortest tool possible for the application as well as using a toolholder with a tighter grip such as a shrink fit. If the problem exists it must be reduce the depth of cut. This will help reduce the force that feeds the chatter and allow to increase the feed rate to compensate the metal removal rate that was lost by using the lower value for the depth of cut. Also, it is recommended to use a better quality end mill as well as a balanced tool which will aid in a more predictable chip thickness and will provide more stability during the machining. Running a lower quality end mill in an unbalanced holder allows for inconsistent chip removal which is a recipe for disaster and a can easily lead to self-excited vibrations. Nevertheless it can be used an adaptive damper. In the following, we present several tests that capture the chatter phenomenon and implicitly its negative effects.

### 2.1 Test setup and used equipment in determining the effects of the chatter

To study the phenomena underlying the stability of machining processes with effects on surface quality, we opted for a distributed and reconfigurable data acquisition and processing system accompanied by a series of stand-alone analysis equipment. A distributed system is a collection of independent computers that appear to the system users as a single computer. Taking this definition into account, the working environment was designed and implemented from a hardware point of view and from here the customized software for the experimental determinations was developed. The various stages in the system generation were tested under both static and dynamic conditions to verify any errors but also to optimize some of the logic elements that were the basis for the software development.

The main component of this measurement chain is the NI CompactRIOTM hardware platform composed of the NI cRIO 9025 real-time controller equipped with an 800 MHz processor, 4 GB nonvolatile memory, 512 MB DDR2 memory and a LabVIEW Real-Time operating system for deterministic control, acquisition and data storage, a reconfigurable 8-port chassis to which an NI cRIO 9234 4-channel C Series input module was attached for dynamic signal acquisition used primarily for high-precision audio frequency measurements. The input channels incorporate IEPE signal conditioners for accelerometers and microphones and simultaneously digitize the signals at rates between 2 and 50 kHz. The software of this platform was developed using the NI LabVIEW® graphical programming environment, the NI LabVIEW Real-Time® module and the NI FPGA® for NI cRIO starting from the specific functions of the specialized libraries within the environment regarding the acquisition and analysis of vibro-acoustic signals (**Figure 8**). It works on three parallel planes, one for data acquisition, one for their saving and the last representing the user interface.

Starting from the performances of the OKUMA M460R-V machining center and the dynamic characteristics of the accelerometers used for vibration measurement, the optimal working parameters for this system were deduced, namely a sampling rate of 51.2 KHz, a data saving rate of 51.2 KS/s for the most faithful reproduction of the process in the post-processing analysis. The system is capable of automatically saving process data at a predefined time, based on the definition of a trigger that can be physical or software (based on the use of a threshold on an input channel) or manually upon user intervention (**Figure 9**).

After the transducers were placed, two sets of measurements were performed, once using the NI cRIO system architecture and once using a PicoScope 4824 PC Oscilloscope. The results obtained were identical in terms of the acquired signals. Once the data acquisition system was validated, a first technology was developed to prepare and execute the experimental plan.

In order to measure the roughness of the manufactured surface on use a surface roughness instrument Mitutoyo MV 2100 (**Figure 10a** [11]). The wear on the rake and clearance surfaces is measured with the help of an digital microscope ISM-PM 200B (**Figure 10b**).

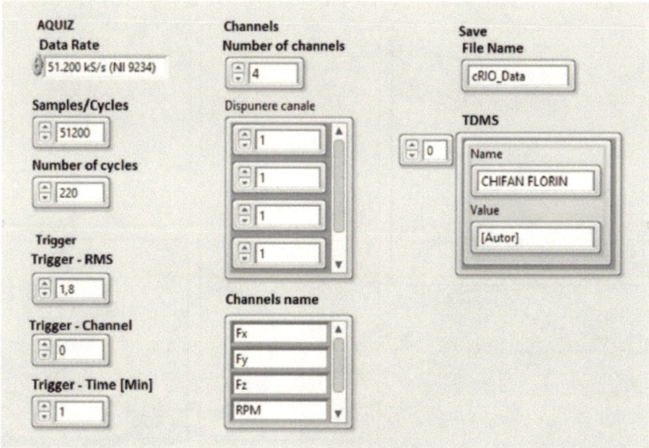


a).

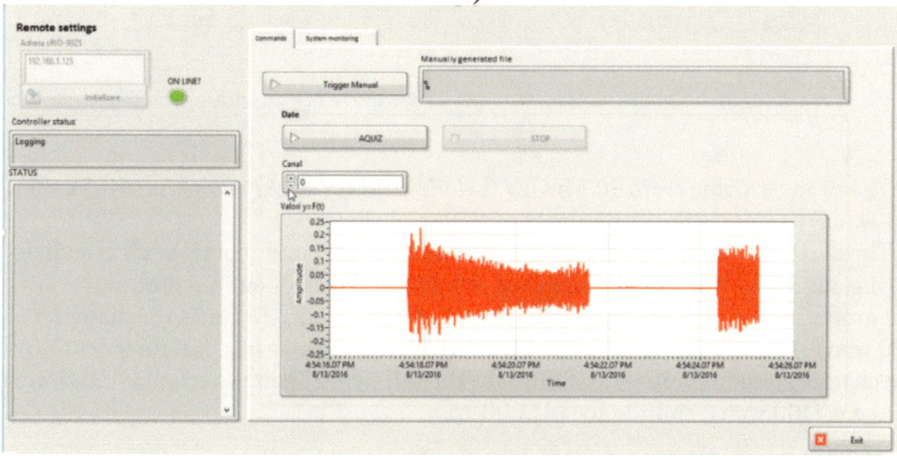


b).

**Figure 8.**
*NI cRIO users interface: (a) window used in defining the working parameters; (b) cutting force data.*

**Figure 9.**
*Okuma M460 R-V machining center and the acquisition system.*

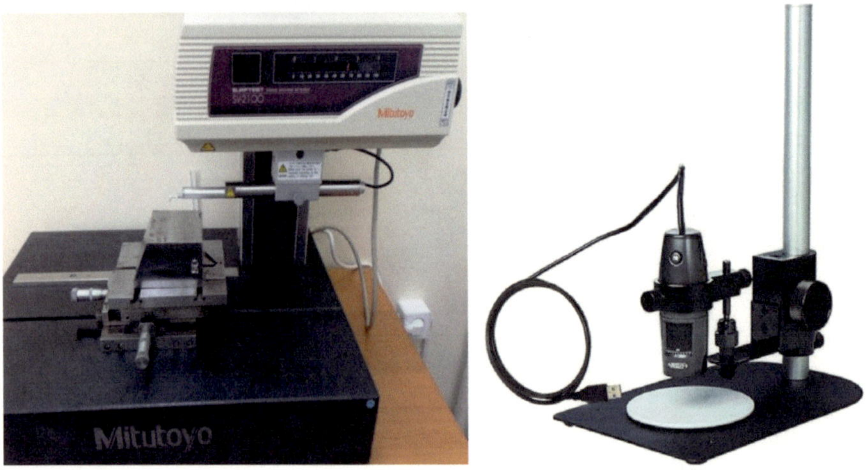


a). b).

**Figure 10.**
*Equipment used in measuring process: (a) surface roughness Instrument MV 2100; (b) digital microscope ISM-PM 200B.*

The probe is made from 90 MnCrV 8 (*DIN*), 90MV 8 (*AFNOR*), 90 MnVCr 8 KU (*UNI*), O2 (*AISI*), with the hardness 40 HRC.

The cutting tool is an end mill with 10 mm diameter and 4 peripheral effective cutting edge. The working regime: feed = 0.2 mm; depth of cut = 6 mm.

I used various overhang for the cutter, the largest being 10 times the diameter (100 mm). The spindle speed varied from 1000 revolutions per minute to 5000 rpm. The chatter phenomenon was capture for the following setup: overhang – 10 times the diameter (100 mm); spindle speed 5000 rpm.

## 2.2 Results

In the following we will present the results. We were able to experiment the chatter phenomenon. This is presented in **Figures 11** and **12**.

The frequency of the vibration is presented for those two cases, with chatter and without, in **Figure 13**.

The self-excited vibrations caused damage to the manufactured surface and also lead to an extensive wear of the cutting tool. This wear increase the frequency of the vibration.

In **Figure 14** is presented the quality of the manufactured surface and in **Figure 15** the measured roughness for these two cases [11].

The chatter produce an excessive wear on the cutting tool both on the clearance surface and on the rake surface. In **Figure 16** we present the geometry of the cutting tool at the beginning of the process and after the chatter vibration.

From the above figures we can see that level of frequency that appears in system is almost six times bigger when we have chatter, 6609 Hz against 1406 Hz. The roughness of the surface of the part is also affected.

## 2.3 Chatter suppression techniques

In order suppress chatter in cutting process we can use one or more measures indicated in **Figure 17** [12].

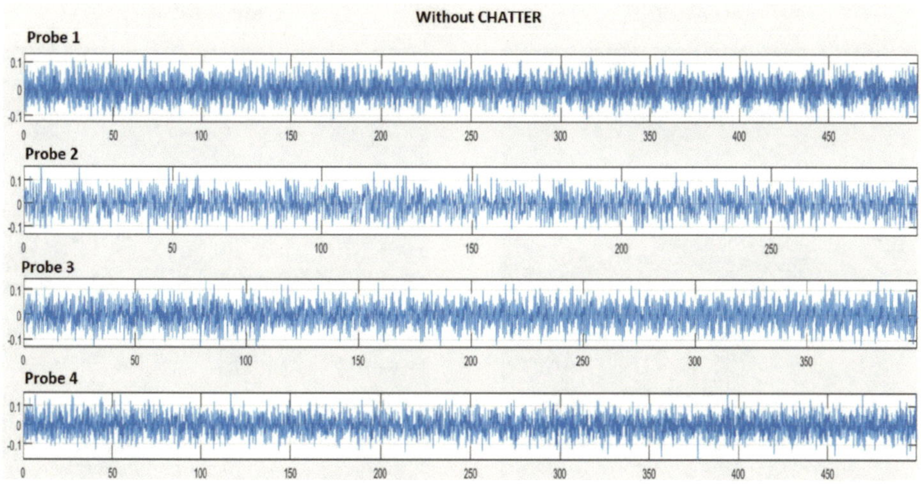


**Figure 11.**
*The dynamicity of the manufacturing process—without chatter.*

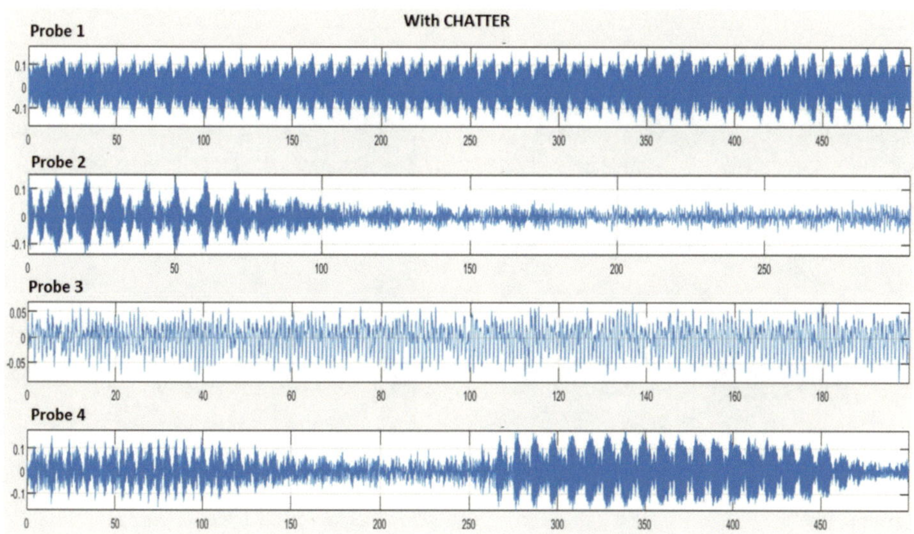


**Figure 12.**
*The dynamicity of the manufacturing process—with chatter.*

The easiest ways to prevent chatter are presented in the following:

1. *Adjusting cutting parameters*:

- Cutting speed: Reducing the cutting speed can help reduce vibration. It is important to adjust according to the material being machined and the type of tool being used.

- Feed per unit length: Too high a feed can cause vibration, so it may be useful to reduce the feed per unit length.

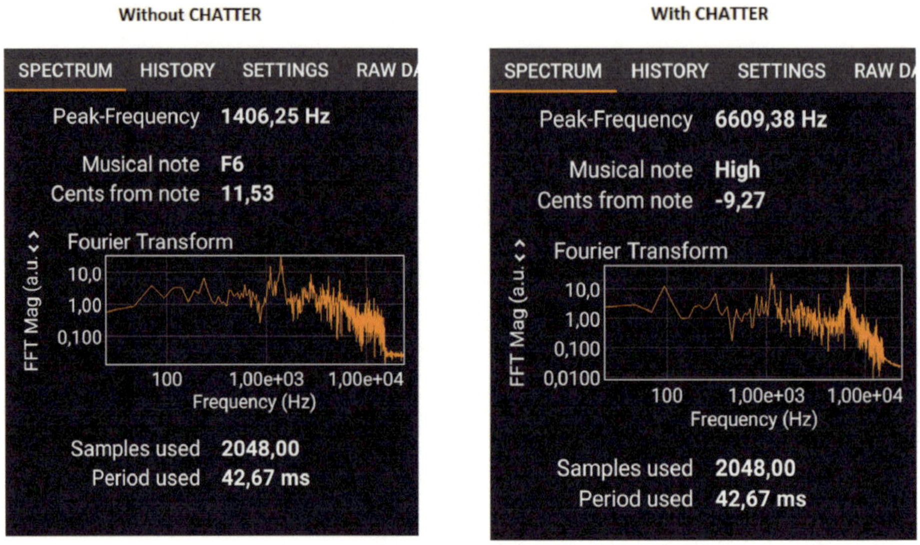


**Figure 13.**
*The frequency of the vibration in cutting process.*

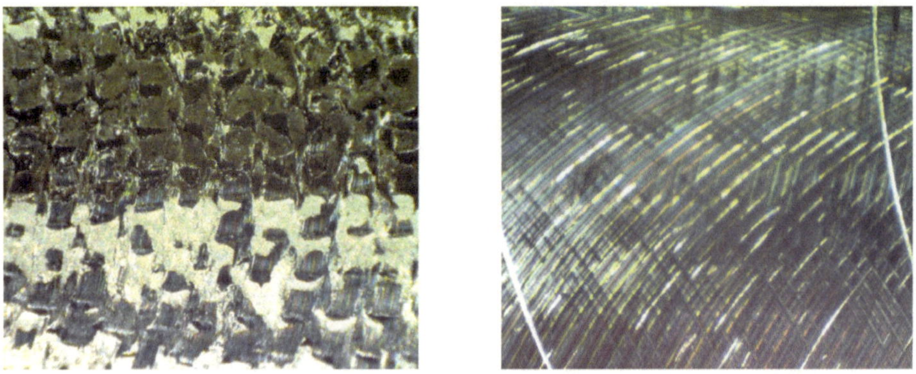

a). b).

**Figure 14.**
*Surface quality: (a) chatter and (b) no chatter.*

- Depth of cut: Reducing the depth of cut or the cutting pitch can help reduce vibration.

2. *Tool and workpiece rigidity*:

- Strengthening the support: Using more rigid supports for the workpiece or tool can help prevent vibration. Ensuring good fixation of the workpiece and tool is crucial.

- Reducing the free length of the tool: The shorter the working length of the tool, the lower the risk of chatter.

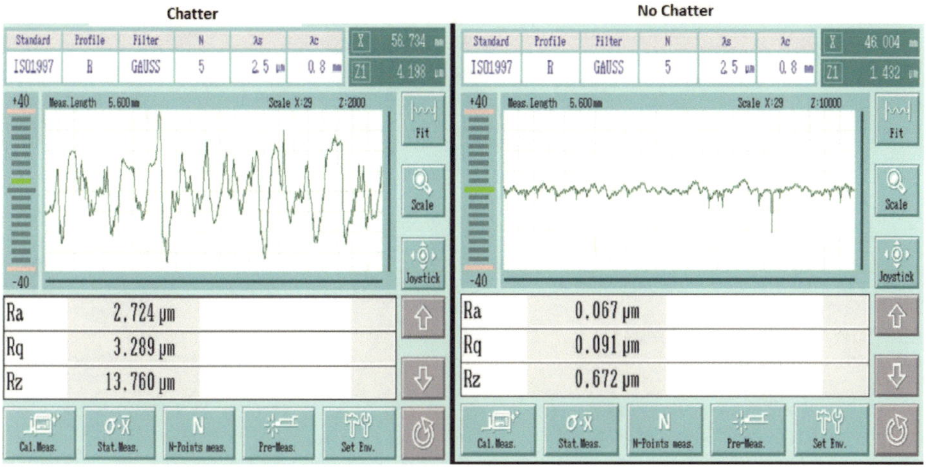


**Figure 15.**
*The measured roughness.*

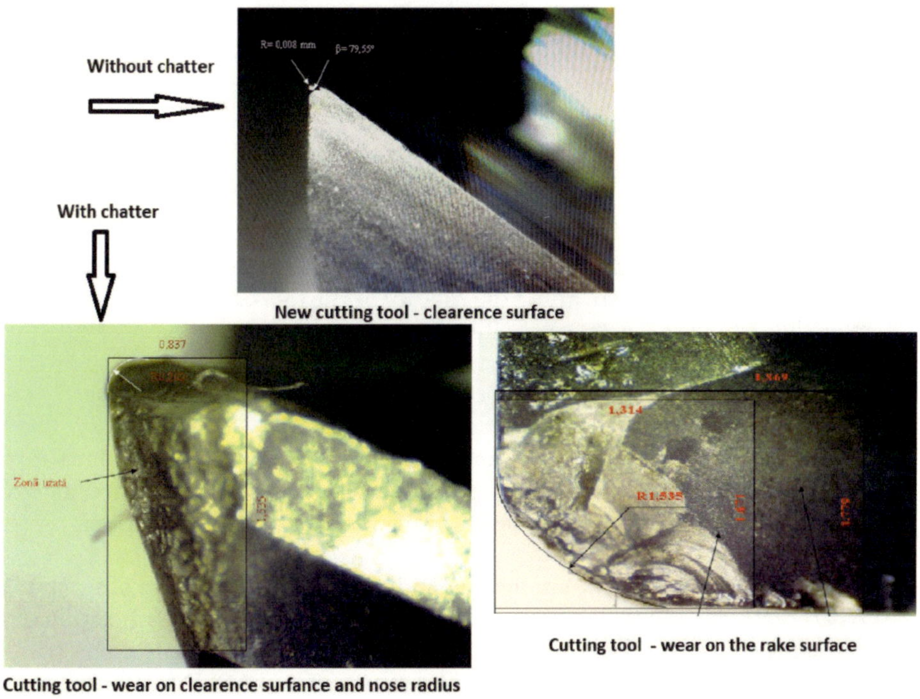


**Figure 16.**
*The wear on the cutting tool.*

3. *Cutting tool geometry*:

- The correct choice of cutting tools can influence vibrations. For example, tools with correctly chosen cutting angles, or tools made of special materials can help reduce vibrations.

**Figure 17.**
*Chatter suppression techniques (adapted from [12]).*

4. *Temperature control*:

 - Ensuring a constant temperature during the cutting process is important, as thermal variations can affect the rigidity of the tool or workpiece and can contribute to the occurrence of vibrations.

5. *Simulations and real-time monitoring:*

 - The use of modern technologies for simulating cutting processes and real-time monitoring of vibrations can help prevent chatter. Sensors and control algorithms can adjust process parameters along the way to avoid vibrations.

Each of these methods can be applied depending on the specific type of machining, the part material, and the equipment available. An appropriate combination of these techniques is usually the most effective method for avoiding chatter.

Another way to solve chatter problems is by constructing of SLD (Stability Lobe Diagram) and changing the process parameters depth of cut (a) and spindle speed (N) accordingly [13]. The SLD is built using a stability model fed with four different inputs: the cutting coefficients or specific forces describing the cutting force, the dynamic parameters of the system, the process parameters and the tool geometry (**Figure 18**).

The qualitative location of the chatter process on the stability diagram SLD is a key factor to select the optimal chatter suppression technique. The relative position of the unstable process is defined by the ratio k between the chatter frequency $f_c$ and tooth passing frequency $f_z$ which depends on the spindle speed N and the number of teeth Z on the tool (Eq. (2)).

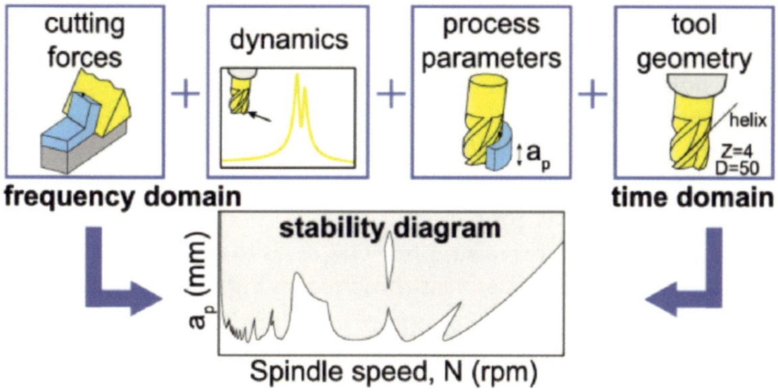


**Figure 18.**
*Stability Lobe Diagram (adapted from [13]).*

Physically, it defines the number of complete waves per period produced by the chatter. It is possible to identify four relative zones according to this ratio (**Figure 19**) [14].

$$k = \frac{f_c}{f_z} = \frac{60 \cdot f_c}{Z \cdot N} \tag{2}$$

*Zone A: process damping zone* ($k > 10$). Process damping is important in this zone, and therefore a high increase in the stability is obtained due to the friction between the flank face with the wavy surface. In this zone, the lower the spindle speed is, the higher the stability boundary is.

*Zone B: intermediate zone* ($10 > k > 3$). The stability limit is close to the absolute stability limit in the whole spindle speed range. This is especially true for high damping values.

*Zone C: high speed zone* ($3 > k > 0.5$). In this zone, the stability can be drastically increased by means of the selection of the spindle speed coincident with one of the stability pockets.

*Zone D: ultra-high speed zone* ($0.5 > k$). The stability can be improved by increasing the spindle speed. The limits of machinability and the power of the spindle in

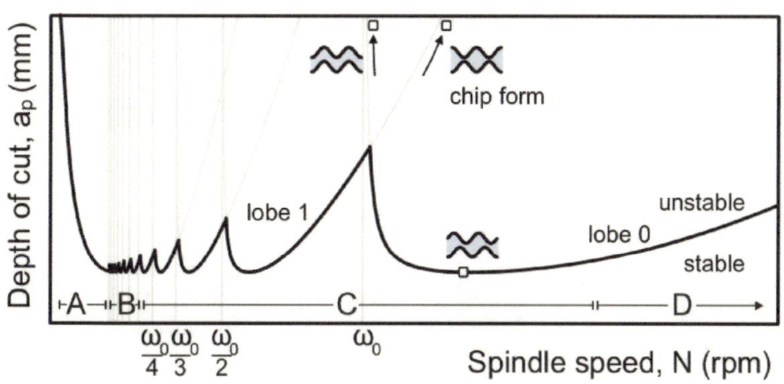


**Figure 19.**
*Relative location on the SLD (adapted from [14]).*

conjunction with the presence of modes at higher frequencies usually limit this otherwise promising option.

It should be noticed that each mode has a spindle speed range where it can create chatter. If the tooth passing frequency is low related to the natural frequency of the mode, the process damping may stabilize the cutting process. On the other hand, if the tooth passing frequency is several times higher than the natural frequency, the mode can hardly create chatter problems.

More sophisticated measures to suppress chatter is to use dampers for tools or for workpiece. These solutions are passive techniques. In **Figure 20** it is presented a solution for boring bars [15].

The anti-vibration mechanism consist of a metallic cylinder of mass m, mounted inside the boring bar. At the ends of this cylinder are two rubber rings that allow the movement of the mass inside the cavity.

The model of the dynamical damper inside of the boring bar is presented in **Figure 21** [8].

The equations that define vibration along the x-axis and y-axis are:

$$m\ddot{x} + c\dot{x} + \left(k_1 + r_l \cdot \cos\beta\right)x + c_1\left(\dot{x} - \dot{x}_1\right) + k_3\left(x - x_1\right) = A\cos\omega t \tag{3}$$

$$m_2\ddot{y} + c\dot{y} + k_2 y + r_l \cdot x\sin\beta = A\sin\omega t \tag{4}$$

$$m_1\ddot{x}_1 - c_1\left(\dot{x} - \dot{x}_1\right) - k_3\left(x - x_1\right) = 0 \tag{5}$$

$$m_2 = m + m_1 \tag{6}$$

The damper for milling process are much more complicated. In **Figure 22** it is presented such a solution [16].

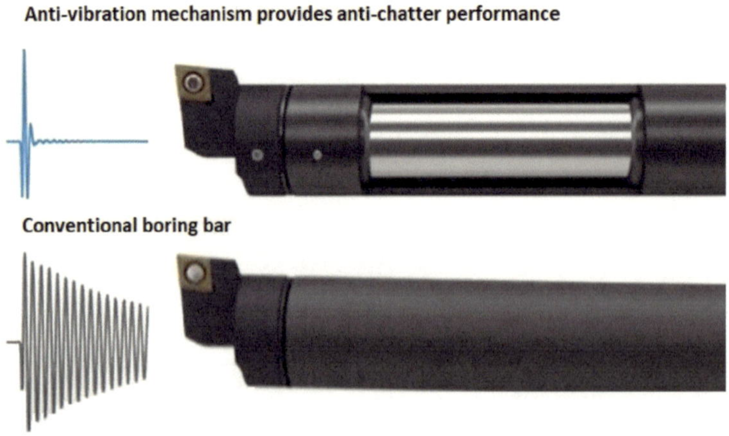


**Figure 20.**
*Damper solution for boring process (adapted from [15]).*

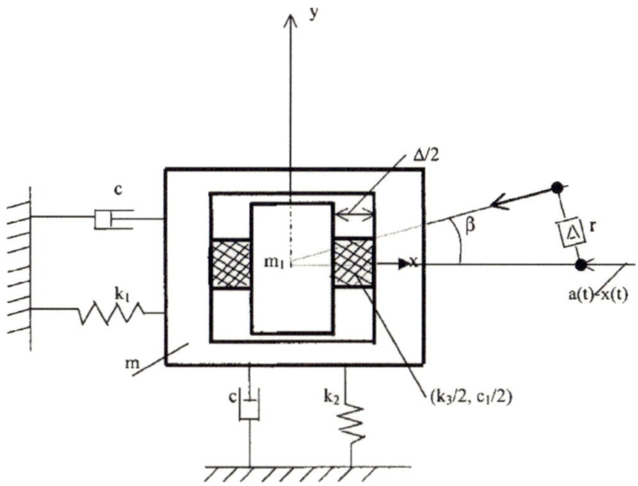


**Figure 21.**
*Model of the dynamical damper inside of the boring bar (adapted from [8]).*

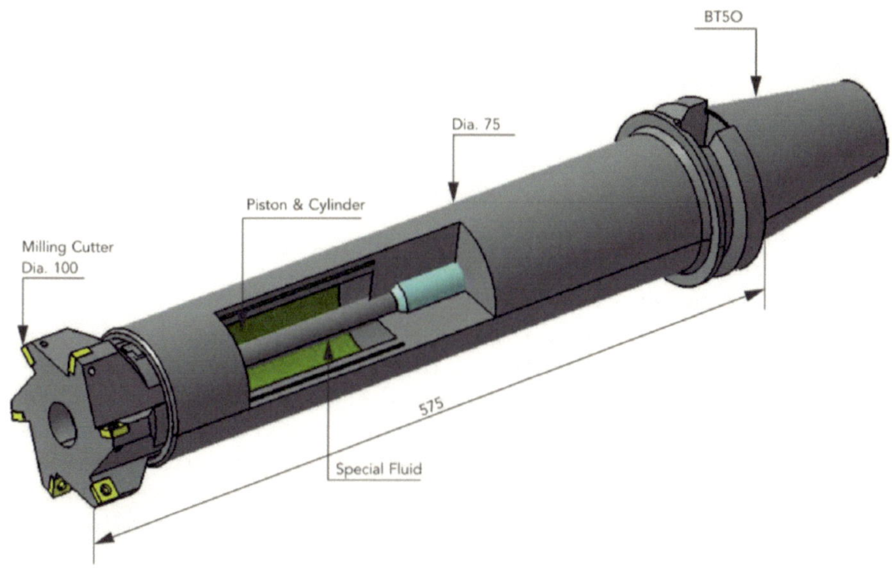


**Figure 22.**
*Damper solution for milling process (adapted from [13]).*

This type of damper use a viscous fluid and the rigidity increase by increasing the pressure inside the damper with the help of a central screw and a piston.

The 2 DOF model of milling is presented in **Figure 23** [17].

Nonlinear differential equations describing the dynamics of the system are presented in the following form:

$$m_1\ddot{x}_1(t)+c_1\dot{x}_1(t)+k_1x_1(t)+k_{11}x_1(t)^3=\sum_{p=1}^{z}F_p(t) \tag{7}$$

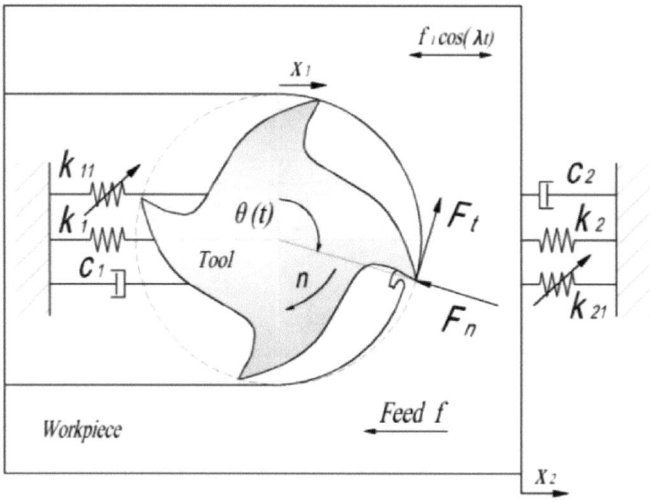


**Figure 23.**
*2 DOF milling model (adapted from [17]).*

$$m_2\ddot{x}_2(t)+c_2\dot{x}_2(t)+k_2x_2(t)+k_{21}x_2(t)^3=-\sum_{p=1}^{z}F_p(t) \tag{8}$$

where:
$m_1$—substitute mass of the tool,
$c_1$—damping of the tool,
$k_1$, $k_{11}$—linear and nonlinear stiffness of the tool,
$m_2$—substitute mass of the workpiece,
$c_2$—damping of the workpiece,
$k_2$, $k_{21}$—linear and nonlinear stiffness of the workpiece.

Active damping devices can add damping to the machine structure or workpiece actively. This technique is also called active structural chatter suppression and has been widely used in heavy-duty milling operations. Different from the chatter resulting from the flexible sources of the machine-workpiece-tool system, chatter in heavy-duty milling processes is generally caused by low-frequency structural modes. Active dampers with proper control strategies can enhance the structural damping without significant changes in the basic structure of the machine [12].

The control methods for active damping devices generally consist of direct velocity feedback (DVF), direct acceleration feedback (DAF), direct position feedback (PF), delayed PF, virtual passive absorber (VPA), proportional integrated derivative (PID), linear quadratic regulators (LQR), linear quadratic Gaussian (LQG), H infinity ($H_\infty$), μ synthesis control, etc. Due to its simplicity and good control performance, DVF is widely used in active damping devices. At present, DVF has become the most efficient control algorithm.

Boring bars have focused a lot of attention due to their inherent flexibility. Piezoelectric actuators were largely used for this application. The piezoelectric actuators are introduced in the body of the boring bar and applied DVF to increase the damping of the bar. Also it can be used a piezo actuator but parallel to the force flow. The vibrations are reduced by a proof-mass absorber generating an inertial force

**Figure 24.**
*Boring bar with magnetorheological (MR) fluid (adapted from [18]).*

that counteracts the disturbance acting in the system. Some solutions use magnetorheological (MR) fluid located around the base of the boring bar to adjust the damping and natural frequency. Later, they optimized the excitation strategy to vary the stiffness of the bar continuously (**Figure 24**) [18].

## 3. Conclusions

Vibration in metal processing by cutting is a complex phenomenon. In the process of mechanical machining by cutting, the occurrence of vibrations is inevitable, and in situations where the amplitude exceeds the limits imposed on the dimensional and shape precision of the generated surfaces, the vibrating phenomenon is harmful.

Due to vibrations in the cutting process, there is wear on the machine tool, premature wear of the cutting tool, especially when using tools with metal carbides or mineral-ceramic inserts, as well as a worsening of the roughness of the manufactured surface, which is unfavorable for finishing operations. Along with the previous shortcomings, vibrations reduce the possibilities of full exploitation of the machine tool and produce noise pollution of the environment.

There are three types of vibrations in metal cutting but self-excited vibration, called chatter, is one of the main limitations in metal removal processes. Chatter may spoil the surface of the part and can also cause large reduction in the life of the different components of the machine tool including cutting tool itself.

If one experience sudden chatter the first thing that must be done is to increase the rigidity of the tool assembly, this can be obtained by using the shortest tool possible for the application as well as using a toolholder with a tighter grip such as a shrink fit. If the problem exists it must be reduce the depth of cut. This will help reduce the force that feeds the chatter and allow to increase the feed rate to compensate the metal removal rate that was lost by using the lower value for the depth of cut. Also, it is recommended to use a better quality end mill as well as a balanced tool which will aid in a more predictable chip thickness and will provide more stability during the machining.

Also there are active or passive suppression technique which were presented in this chapter.

## Author details

Cătălin Gabriel Dumitraş*, Mihaiţă Horodincă, Adriana Munteanu, Florin Chifan, Dragoş-Florin Chitariu and George-Gabriel Chiriac
Faculty of Machine Manufacturing and Industrial Management, Technical University Gheorghe Asachi, Iasi, Romania

*Address all correspondence to: catalin-gabriel.dumitras@academic.tuiasi.ro

## References

[1] Koudinov VA. Dinamica mașinilor-unelte (Machine tool dynamics). București: Editura tehnica; 1970. 386 p

[2] Moraru V, Ispas C, Rusu S. Vibrațiile și stabilitatea mașinilor-unelte (Machine tool vibrations and stability). București: Editura Tehnica; 1982. 350 p

[3] Deacu L, Pavel G. Vibrații la mașini-unelte. (Machine tool vibrations). Cluj - Napoca: Editura Dacia; 1977

[4] Tobias AS. Machine-Tool Vibration. New York: John Wiley and Sons; 1965. 380 p. ISBN 9780216874541

[5] Ispas C, Simion FP. Vibratiile mașinilor-unelte. Teorie și aplicatii. (Machine tool vibration. Theory and Applications.). București: Editura Academiei Romane; 1986

[6] Ştețiu G, Lăzărescu I, Oprean C-tin, Ştețiu M. Teoria şi Practica Sculelor aşchietoare (Theory and Practice of Cutting Tools). Vol. I. Sibiu, Romania: Universității din Sibiu; 1994. 753 p

[7] Harris MC, Piersol AG. Harris' Shock and Vibrations Handbook. 5th ed. New York: McGraw-Hill; 2002. 1456 p. ISBN 0-07-137081-1

[8] Crivacucea CO, Probleme privind amortizarea vibrațiilor produse în procesul prelucrărilor mecanice (Problems regarding the damping of vibrations produced in the mechanical cutting process) [thesis], Universitatea "Politehnica" Timisoara. 2001

[9] Korka ZI, Miclosina CO, Cojocaru V. An experimental study of the cutting forces in metal turning. Vol. XX, no. 2. Resita, Romania: "Eftimie Murgu" University Annals; 2013. pp. 25-32. ISSN 1453-7397

[10] Kalendjeridze VG. Influenta proprietăților mecanice ale metalelor asupra aparitiei vibrațiilor la strunjire. (The influence of the mechanical properties of metals on the occurrence of vibrations during turning). In: Stanki I instrument. No. 7. Mashgiz, Moskva, Russia; 1964

[11] Horodincă M, Chifan F, Păduraru E, Dumitraș CG, Munteanu A, Chitariu DF. A study of 2D roughness periodical profile on a flat surface generated by milling with a ball nose mill. Materials. 2024;**17**(6):1425. DOI: 10.3390/ma17061425

[12] Li Z, Song Q, Jin P, Liu Z, Wang B, Ma H. Chatter suppression techniques in milling processes: A state of the art review. Chinese Journal of Aeronautics. 2024;**37**(7):1-23. DOI: 10.1016/j.cja.2023.10.001

[13] Tobias AS, Fishwick W. Theory of regenerative machine tool chatter. The Engineer. 1958. pp. 199-203

[14] Smith S, Tlusty J. Current trends in high-speed machining. Journal of Manufacturing Science and Engineering. 1997;**119**(4B):664-666. DOI: 10.1115/1.2836806

[15] Kyocera. KAV – Interchangeable Head Boring Bars with Anti-vibration Dampener System. 2024. Available from: https://www.youtube.com/watch?v=xSq7AvjW7UI; https://www.kyocera-unimerco.com/en-global/about/news/kav [Accessed: January 17, 2025]

[16] Sigma toolings. SAMURAI Anti-Vibration Tool. 2024. Available from: https://sigmatoolings.com/samurai-tool/ [Accessed: January 17, 2025]

[17] Weremczuka A, Rusineka R, Warminskia J. The concept of active elimination of vibrations in milling process. In: 15th CIRP Conference on Modelling of Machining Operations, Procedia CIRP. Vol. 31. Elsevier; 2015. pp. 82-87. DOI: 10.1016/j.procir.2015.03.036

[18] Munoa J, Beudaert X, Dombovarib Z, Altintas Y, Budak E, Brecher C, et al. Chatter suppression techniques in metal cutting. CIRP Annals Manufacturing Technology. 2016;**65**:785-808. DOI: 10.1016/j.cirp.2016.06.004

**Chapter 4**

# Frequency Based Substructuring: Coupling and Decoupling of Dynamic Systems

*Matthew Clontz*


## Abstract

Dynamic substructuring (DS) is a process by which the dynamics of a complex system are analyzed by breaking the system down into separate components and analyzed individually. The dynamic information for the components can then be coupled together to form the assembled system, or a component can be decoupled from the assembled system. The standard formulation for DS in frequency domain was first developed in the 1980s. Due to limitations in measurement and computational power, interest in the field declined. With the increase in experimental and computational capabilities which became available by the early 2000s, the structural dynamic community showed a renewed interest in DS techniques. Since that time, a variety of improved formulations for both coupling and decoupling have been investigated. Additionally, the emphasis of DS in experimental applications has driven research into difficulties and subsequent improvements to experimental practices (i.e. for rotations, error propagation, etc.). This chapter presents the fundamental theory and mathematical background of dynamic substructuring in the frequency domain (referred to as Frequency Based Substructuring).


**Keywords:** dynamic substructuring, frequency-based substructuring, coupling, decoupling, frequency-based substructuring derivation

## 1. Introduction

Dynamic substructuring (DS) is a process by which the dynamics of a complex system are analyzed by breaking the system down into separate components and analyzed individually. The dynamic information for the components can then be coupled together to form the assembled system, or a component can be decoupled from the assembled system. The standard formulation for DS in frequency domain was first developed in the 1980s. Due to limitations in measurement and computational power, interest in the field declined. With the increase in experimental and computational capabilities which became available by the early 2000s, the structural dynamic community showed a renewed interest in DS techniques. Since that time, a variety of improved formulations for both coupling and decoupling have been investigated. Additionally, the emphasis of DS in experimental applications has driven research into difficulties and subsequent improvements to experimental practices (i.e. for rotations, error propagation, etc.). This chapter presents the fundamental theory and mathematical background of dynamic substructuring in the frequency domain.

The concept of breaking down a complex problem into multiple, but simple problems is not a new idea. It has been used throughout history in widely varying fields ranging from military applications to mathematics and engineering. Dynamic substructuring is a process that utilizes the concept of breaking down a complex mechanical system into components or "substructures." There are several main advantages or reasons that one might use dynamic substructuring in place of global system analysis, such as: the ability to analyze large or complex systems on a component basis, ease of isolating and identifying local dynamic behavior, and the ability to combine dynamic information from different sources (i.e. FEA and experimental) [1].

In general, dynamic substructuring can be thought of in terms of either coupling or decoupling substructures in the physical, modal or frequency domains. The next section will give a brief overview of the theory associated with dynamic substructuring coupling or decoupling in each domain. While all the same information is theoretically present in each domain, analyses in each are carried out differently and each have different advantages. Subsequently, the remainder of the chapter will focus on substructuring in the frequency domain.

Dynamic substructuring in the frequency domain is often called "frequency based substructuring" (FBS). The classical FBS method was developed in [2]. A decade later, the basic theory of FBS was well established, but the method was often troubled by experimental difficulties. Additionally, with the limit in computation power available in the late twentieth century, many complex structures of interest were unable to be modeled numerically. However, due to an increase in computing power and testing capabilities available in the twenty-first century, there has been a renewed interest in FBS since the mid 2000's [3].

After a brief overview of dynamic substructuring in the three domains, the remainder of the chapter will focus on coupling or decoupling procedures in the frequency domain. First, the remaining sections of the present chapter will provide a brief history of FBS methods and the associated mathematical background.

## 2. Substructuring in the physical domain

As mentioned previously, dynamic substructuring can be formulated in physical, modal, and frequency domains. Although each domain theoretically contains the same mathematical information, the formulation in each can be quite different in practice. Corresponding to the differences in application, formulations in each domain have distinct differences and advantages. The remainder of this section will give a brief overview of the formulation in each domain as presented by De Klerk et al. [1]. **Figure 1** shows the general relationship between the different domains.

This figure also indirectly shows certain limitations that arise between substructuring in the various domains. For instance, when moving to the modal domain, certain information may be inherently lost due to the need to use a finite number of modes. This loss of information and other issues with the specific domains will be addressed in the corresponding parts of this section.

Before dynamic substructuring in each specific domain can be discussed, several important details must first be presented. By definition, a substructure is a structure that interacts with other adjacent structures within some assembly or system.

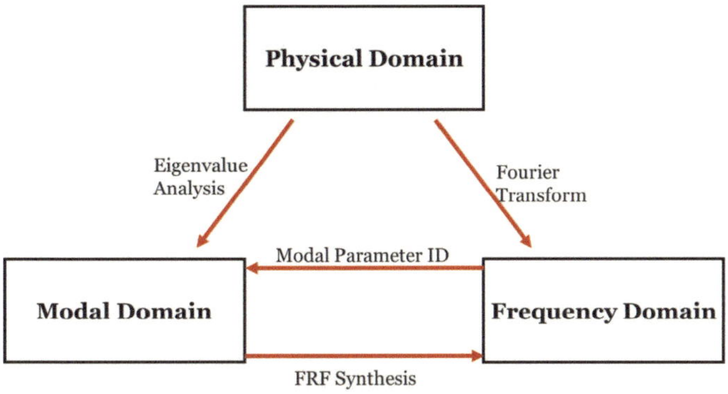


**Figure 1.**
*Relationship of system dynamics between the three domains.*

Regardless of what method is used to couple (or decouple) a structure, there are two conditions which must always be met between coupled structures:

1. The "*compatibility condition*" requires that any coupled substructures have compatible displacements at the interface. This condition is equivalent to two substructures remaining joined at their connection or "interface point," which intuitively makes sense in a physical perspective.

2. The "*equilibrium condition*" requires that the forces on the substructure interfaces to be in equilibrium. Again, this condition makes sense in a physical perspective as it corresponds to a force balance for the given interface point.

Generally, there are two types of assembly methods to determine the characteristics of the coupled system. These assembly methods are termed "primal" or "dual" depending on the choice of unknowns. A primal assembly is one in which displacements are treated as unknown, while a dual assembly is one in which forces are treated as unknown. There are also assembly methods that treat both forces and displacements as unknowns either separately or in some combination. Primal and dual methods can be used in each domain and will be discussed the following subsections.

## 2.1 Physical domain

A structure is described in the physical domain by its mass, stiffness, and damping characteristics. In the typical way, the equations of motion of a discretized dynamic subsystem (s) can be written as:

$$M^{(s)}\ddot{x}^{(s)}(t) + C^{(s)}\dot{x}^{(s)}(t) + K^{(s)}x^{(s)}(t) = f^{(s)}(t) + g^{(s)}(t) \quad (1)$$

Where M(s), C(s), and K(s) are mass, damping, and stiffness matrices respectively; x(s) is a vector of displacements (or degrees of freedom), f(s) is a vector of external forces, and g(s) is a vector of connecting forces with other substructures. It is important to note that Eq. (1) is for one specific substructure of an assembly, (s). It is also important to note that the system described in Eq. (1) is assumed to be a linear, time-invariant (LTI) system. Within this framework, the external force vector can be viewed as constraining forces for the previously mentioned necessary conditions (compatibility and equilibrium).

There have been methods developed for nonlinear or time variant systems in recent years, but no standard set of methods has been established yet [4, 5]. The lack of a standard, nonlinear method can be attributed to the difficulty of generalizing nonlinear systems. As such, the present work will focus on the well-established, linear system methods.

In the more general case, where there are n substructures that are coupled in an assembly, the system can be written as:

$$M\ddot{x} + C\dot{x} + Kx = f + g \tag{2}$$

Where M, C, and K are now block diagonal matrices, while f, g, and x are now columns of vectors consisting of components from each substructure. Note that f, g, and x are still time dependent though the notation has been dropped in the interest of simplicity. The terms of Eq. (2) can be written more specifically as follows:

$$M \triangleq diag\left(M^{(1)}, M^{(2)}, \dots, M^{(n)}\right) = \begin{bmatrix} M^{(1)} & \cdots & \cdot \\ \vdots & \ddots & \vdots \\ \cdot & \cdots & M^{(n)} \end{bmatrix}$$

$$\boldsymbol{C} \triangleq \boldsymbol{diag}\left(\boldsymbol{C}^{(1)}, \boldsymbol{C}^{(2)}, \dots, \boldsymbol{C}^{(n)}\right)$$

and

$$\boldsymbol{K} \triangleq \boldsymbol{diag}\left(\boldsymbol{K}^{(1)}, \boldsymbol{K}^{(2)}, \dots, \boldsymbol{K}^{(n)}\right)$$

$$\boldsymbol{x} = \begin{bmatrix} \boldsymbol{x}^{(1)} \\ \vdots \\ \boldsymbol{x}^{(n)} \end{bmatrix}, \boldsymbol{f} = \begin{bmatrix} \boldsymbol{f}^{(1)} \\ \vdots \\ \boldsymbol{f}^{(n)} \end{bmatrix}, \boldsymbol{g} = \begin{bmatrix} \boldsymbol{g}^{(1)} \\ \vdots \\ \boldsymbol{g}^{(n)} \end{bmatrix}$$

From the expressions for the block diagonal matrices, it is important to note that all of the off-diagonal terms are zero. In this context "zero" means that the off-diagonal terms are matrices of corresponding size and full of zeros.

With the generalized form of the equations of motion as shown in Eq. (2), the compatibility condition is expressed as:

$$Bx = 0 \tag{3}$$

Where B is a signed Boolean matrix which operates on the interface degrees of freedom (i.e. DOFs involved in coupling). The matrix B is essentially a matrix which denotes the connections between the entire set of DOFs. A more comprehensive discussion of this matrix is given in [1]. However, it is worth noting here that B may not be Boolean if certain other conditions are applied to the interfaces. These non-Boolean cases are beyond the scope of the present work, but the following discussion is still valid for those cases.

The equilibrium condition can similarly be written as:

$$L^T g = 0 \tag{4}$$

Where L is a different Boolean matrix designed to localize the interface DOFs of each substructure from the global, dual set of DOFs. Eq. (4) states that the resultant force must be zero when the dual connection forces are summed at a given interface point. In reality, the equilibrium condition is being applied for a discretized system and thus will result in an “equilibrium error” when applied to the actual continuous system. For the purposes of this work, Eq. (4) will be treated as enforcing exact equilibrium.

Continuing with the brief overview of the physical domain formulation [1], the complete system of equations can be written as:

$$\begin{cases} M\ddot{x} + C\dot{x} + Kx = f + g \\ Bx = 0 \\ L^T g = 0 \end{cases} \tag{5}$$

Recall that the system of Eq. (5) is for any number of structures (this number was denoted as “n”). Using Eq. (5), a formulation for the coupled structure can be obtained in either a primal or dual manner.

### *2.1.1 Physical domain: Primal formulation*

In a primal formulation, interfaces forces are eliminated as unknowns, thus allowing the system to be coupled by using the compatibility condition. A unique set of interface degrees of freedom must be defined, which is expressed mathematically as:

$$x = Lq \tag{6}$$

Where $\boldsymbol{x}$ and $\boldsymbol{L}$ are as previously defined, and q is the unique set of interface DOFs. By definition, Eq. (6) indicates that the substructure DOFs ($\boldsymbol{x}$) are obtained using the unique set which, in turn, implies that the compatibility condition will be satisfied for any set $\boldsymbol{q}$. This can be understood mathematically by:

Where $\boldsymbol{x}$ and $\boldsymbol{L}$ are as previously defined, and q is the unique set of interface DOFs. By definition, Eq. (6) indicates that the substructure DOFs ($\boldsymbol{x}$) are obtained using the unique set which, in turn, implies that the compatibility condition will be satisfied for any set $\boldsymbol{q}$. This can be understood mathematically by:

$$Bx = BLq = 0 \forall q \tag{7}$$

This relationship implies that L is actually the null space of B or vice versa. One useful result of this property is that only one Boolean matrix needs to be formulated in the assembly process of calculation the response of the coupled system. For more details on this property see De Klerk et al. [1]

Since the choice of the unique set $\boldsymbol{q}$ automatically satisfies the compatibility condition in Eqs. (5), (6) can thus be substituted into Eq. (5) to obtain a reduced system:

$$\begin{cases} ML\ddot{q} + CL\dot{q} + KLq = f + g \\ L^T g = 0 \end{cases} \tag{8}$$

The $\boldsymbol{g}$ term can be eliminated from the equations of motion by pre-multiplying by $\boldsymbol{L}^T$. A new set of system matrices can be assembled to simplify the primal assembled system:

$$\begin{cases} M \triangleq L^T ML \\ C \triangleq L^T CL \\ K \triangleq L^T KL \\ f \triangleq L^T fL \end{cases} \tag{9}$$

The above set of matrices completes the reduction of the original system from Eq. (5) to:

$$\tilde{M}\ddot{q} + \tilde{C}\dot{q} + \tilde{K}q = \tilde{f} \tag{10}$$

The concepts which were applied to obtain Eq. (10) will later be used again to obtain a similar primal formulation in the frequency domain.

### 2.1.2 Physical domain: Dual formulation

In contrast to the primal formulation, the dual formulation eliminates the displacements as unknowns and the assembly is solved using the equilibrium condition. Corresponding to the choice of eliminating displacements as unknowns, the dual formulation retains the full set of global DOFs. This statement is best explained by a walk-through of the dual assembly procedure. First, interface forces are chosen in the form:

$$g = -B^T\lambda \tag{11}$$

Where $\lambda$ are Lagrange multipliers which correspond physically to the interface for intensities. Due to the construction of the Boolean matrix $\boldsymbol{B}$, this choice of form for the interface forces ensures that each force will be balanced by a corresponding force in the opposite direction for any given interface DOF.

The equilibrium condition can then be rewritten as follows.

$$L^T g = -L^T B^T \lambda = 0 \tag{12}$$

Recall that it was previously stated that $\boldsymbol{L}$ and $\boldsymbol{B}$ were actually null spaces for each other which implies that the above condition is automatically satisfied. As in the primal formulation, the system of Eq. (5) can be reduced to the following:

$$\begin{cases} M\ddot{x} + C\dot{x} + Kx + B^T\lambda = f \\ Bx = 0 \end{cases} \tag{13}$$

This expression of the dual formulation in the physical domain can finally be shown in the following block matrix representation:

$$\begin{bmatrix} M & 0 \\ 0 & 0 \end{bmatrix}\begin{bmatrix} \ddot{x} \\ \lambda \end{bmatrix} + \begin{bmatrix} C & 0 \\ 0 & 0 \end{bmatrix}\begin{bmatrix} \dot{x} \\ \lambda \end{bmatrix} + \begin{bmatrix} K & B^T \\ B & 0 \end{bmatrix}\begin{bmatrix} x \\ \lambda \end{bmatrix} = \begin{bmatrix} f \\ 0 \end{bmatrix} \tag{14}$$

This concludes the discussion of dynamic substructuring or decoupling in the physical domain. From an application perspective, discussion of substructuring in the physical domain is primarily of use as a description of the processes and principles required. In practice, values to populate the mass, damping, and stiffness matrices are not generally known. Instead, it is common to obtain information about the physical systems, namely receptances or mobilities, through modal testing and analysis [6]. With that application in mind, the remainder of the chapter will focus on frequency based substructuring (FBS). The following section includes a brief history of the FBS method and its extensions as well as a mathematical background of said methods.

## 3. Substructuring in the frequency domain - history and mathematical background of FBS methods

The FBS method in its currently recognized form was first presented in the 1980's [2]. The method was described as "a generalized frequency domain substructure synthesis methodology" with the application of using the new method to analyze helicopter airframe vibration. Before the FBS methodology is discussed, the necessary mathematical background and notation will be presented. Other names for the FBS method include: impedance based substructuring, receptance coupling (or modelling), and admittance coupling (or modelling).

### 3.1 FBS background math and notation

Recall from the previous section that Eq. (15) describes the equations of motion of the given system in the physical domain. Eq. (15) is repeated here for convenience:

$$M\ddot{x} + C\dot{x} + Kx = f + g \tag{15}$$

As in the original FBS method proposed in [2], the interface force vector, $\boldsymbol{g}$, is omitted in the following mathematical discussion. With $\boldsymbol{g}$ omitted, Eq. (2) can be transformed into the frequency domain using a Fourier transform:

$$\left[-\omega^2\boldsymbol{M} + j\omega\boldsymbol{C} + \boldsymbol{K}\right]\boldsymbol{x}(\omega) = \boldsymbol{f}(\omega) \tag{16}$$

Where the left-hand side of Eq. (16) defines the dynamic stiffness matrix, $\boldsymbol{Z}$, so that the system may be written in a condensed form as:

$$\boldsymbol{Z}(\omega)\boldsymbol{x}(\omega) = \boldsymbol{f}(\omega) \tag{17}$$

It is worth noting that the impedance matrix, $\boldsymbol{Z}$, is also often referred to as the dynamic stiffness. An additional matrix can be defined as:

$$\boldsymbol{Y} = \boldsymbol{Z}^{-1} \tag{18}$$

Here, the notation of frequency dependence has been dropped in the interest of clarity. The matrix $\boldsymbol{Y}$ is often referred to as the receptance (or admittance) matrix [6]. Alternatively, for velocity and acceleration, similar matrices are termed mobility or accelerance, respectively. Though some differences exist between substructuring with receptance, mobility, and accelerance matrices, the following derivation for a simplified system holds for all three. Eq. (17) can be rewritten using the mobility matrix in place of the impedance so that it appears as:

$$x = Yf \tag{19}$$

It is useful to view the system as having the form of Eqs. (17) or (19) as opposed to the expanded form in Eq. (16). Mass, stiffness, and damping matrices are not generally known or easily formulated for complex systems. On the other hand, it is much easier and there are well established methods for obtaining receptances or mobilities through modal testing [6].

### 3.2 Simplified FBS formulation

Many academic papers and even textbooks that address dynamic substructuring skip over portions of the background math and derivations that may be non-obvious to a given reader. In the interest of being thorough, that background math will be presented here. **Figure 2** displays the simplest case of coupling between two structures in which the structures are couple by a single DOF.

In **Figure 2**, A and B represent two arbitrary structures that are coupled by a single coordinate to form a new structure that will be called C. In the form of Eq. (19):

$$\boldsymbol{x}_A = \boldsymbol{Y}_A \boldsymbol{f}_A \tag{20}$$

$$\boldsymbol{x}_B = \boldsymbol{Y}_B \boldsymbol{f}_B \tag{21}$$

The compatibility and equilibrium conditions can be applied at the connection point between substructures A and B:

$$\boldsymbol{x}_C = \boldsymbol{x}_A = \boldsymbol{x}_B \tag{22}$$

$$\boldsymbol{F}_C = \boldsymbol{F}_A + \boldsymbol{F}_B \tag{23}$$

Which yields the following after some manipulation:

$$\boldsymbol{Y}_C^{-1} = \boldsymbol{Y}_A^{-1} + \boldsymbol{Y}_B^{-1} \tag{24}$$

Equation (24) can be extended to a system in which there are several DOFs involved in the coupling procedure. However, no additional internal DOFs are included in this analysis. Conceptually, Eq. (24) will give the FRF properties of the coupled system as a combination of the two independent subsystems. It is important

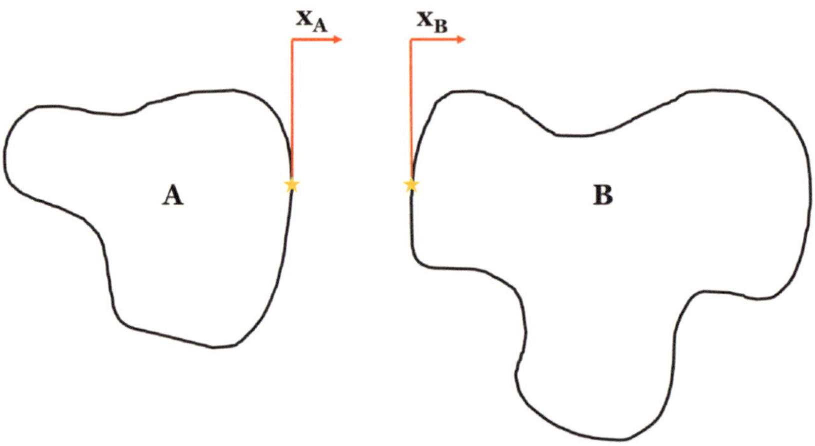


**Figure 2.**
*Arbitrary structures, A and B, coupled by a single coordinate, $x_A$ and $x_B$, on each structure.*

to note that Eq. (24) could be written in terms of dynamic stiffnesses rather than receptances (i.e. using $\boldsymbol{Z}$ as opposed to $\boldsymbol{Y}^{-1}$). For the following analysis, the receptances will be used rather than dynamic stiffnesses, but when the analysis is extended to include multiple DOFs, which may or may not be involved in the coupling, the dynamic stiffness method becomes much more straightforward algebraically.

As the analysis is continued in terms of receptances, there is a major limitation of the form of Eq. (24) that must be addressed. Even for this simplified representation, the current form requires three matrix inversions. Matrix inverse operations are computationally expensive and can end up magnifying errors within the data as they are carried out. The form of Eq. (24) can be changed through the following steps, as derived in [6], to reduce the number of necessary matrix inversions.

$$\begin{aligned}\boldsymbol{Y}_C^{-1} &= \boldsymbol{Y}_A^{-1} + \boldsymbol{Y}_B^{-1} \\ &= \boldsymbol{Y}_A^{-1}(\boldsymbol{I} + \boldsymbol{Y}_A\boldsymbol{Y}_B^{-1}) \\ &= \boldsymbol{Y}_A^{-1}(\boldsymbol{Y}_B + \boldsymbol{Y}_A)\boldsymbol{Y}_B^{-1}\end{aligned}$$

Which, after inversion, yields:

$$\boldsymbol{Y}_C = \boldsymbol{Y}_B(\boldsymbol{Y}_B + \boldsymbol{Y}_A)^{-1}\boldsymbol{Y}_A \tag{25}$$

Equation (25) is much more efficient than Eq. (24) from a computation standpoint. Equation (25) can be rearranged to obtain yet another version that is more useful for modification purposes:

$$\begin{aligned}\boldsymbol{Y}_C &= \boldsymbol{Y}_B(\boldsymbol{Y}_B + \boldsymbol{Y}_A)^{-1}\boldsymbol{Y}_A \\ &= (\boldsymbol{Y}_A + \boldsymbol{Y}_B - \boldsymbol{Y}_A)(\boldsymbol{Y}_B + \boldsymbol{Y}_A)^{-1}\boldsymbol{Y}_A \\ &= (\boldsymbol{Y}_A + \boldsymbol{Y}_B)(\boldsymbol{Y}_B + \boldsymbol{Y}_A)^{-1}\boldsymbol{Y}_A - \boldsymbol{Y}_A(\boldsymbol{Y}_B + \boldsymbol{Y}_A)^{-1}\boldsymbol{Y}_A \\ &= \boldsymbol{I}\boldsymbol{Y}_A - \boldsymbol{Y}_A(\boldsymbol{Y}_B + \boldsymbol{Y}_A)^{-1}\boldsymbol{Y}_A\end{aligned}$$

From which the final desired form is obtained as:

$$\boldsymbol{Y}_C = \boldsymbol{Y}_A - \boldsymbol{Y}_A(\boldsymbol{Y}_B + \boldsymbol{Y}_A)^{-1}\boldsymbol{Y}_A \tag{26}$$

Which is considered the most useful form for structural modification purposes [6].

### 3.3 Generalized FBS formulation

The preceding analysis for the simplified system can readily be extended to the more general case of subsystems with arbitrarily large numbers of DOFs and coupled at any number of those DOFs. In such a case the receptance matrix can be partitioned into portions which correspond to the coupling DOFs and those that do not:

$$\boldsymbol{Y}_A = \begin{bmatrix} \boldsymbol{Y}_{ii}^A & \boldsymbol{Y}_{ic}^A \\ \boldsymbol{Y}_{ci}^A & \boldsymbol{Y}_{cc}^A \end{bmatrix} \tag{27}$$

$$\boldsymbol{Y}_B = \begin{bmatrix} \boldsymbol{Y}_{ii}^B & \boldsymbol{Y}_{ic}^B \\ \boldsymbol{Y}_{ci}^B & \boldsymbol{Y}_{cc}^B \end{bmatrix} \tag{28}$$

Where the subscripts denote the specific receptance and the superscript denotes the substructure from which the receptance originates. For example,

- $\boldsymbol{Y}_{ii}^A$ denotes the partition of the receptance matrix for the internal (uncoupled) DOFs on substructure A
- $\boldsymbol{Y}_{ic}^A$ denotes the cross-partition of the interface/internal (i.e. coupled and uncoupled) receptances
- $\boldsymbol{Y}_{cc}^A$ denotes the partition of the interface (coupling) receptances from substructure A.

It is worth noting that there are corresponding impedance matrices ($\boldsymbol{Z}$) for each receptance. Using the receptances as defined in Eqs. (27) and (28), an expanded form of Eq. (19) is as follows:

$$\begin{Bmatrix} \boldsymbol{x}_i^A \\ \boldsymbol{x}_c^A \end{Bmatrix} = \begin{bmatrix} \boldsymbol{Y}_{ii}^A & \boldsymbol{Y}_{ic}^A \\ \boldsymbol{Y}_{ci}^A & \boldsymbol{Y}_{cc}^A \end{bmatrix} \begin{Bmatrix} \boldsymbol{f}_i^A \\ \boldsymbol{f}_c^A \end{Bmatrix} \tag{29}$$

$$\begin{Bmatrix} \boldsymbol{x}_i^B \\ \boldsymbol{x}_c^B \end{Bmatrix} = \begin{bmatrix} \boldsymbol{Y}_{ii}^B & \boldsymbol{Y}_{ic}^B \\ \boldsymbol{Y}_{ci}^B & \boldsymbol{Y}_{cc}^B \end{bmatrix} \begin{Bmatrix} \boldsymbol{f}_i^B \\ \boldsymbol{f}_c^B \end{Bmatrix} \tag{30}$$

With the formulation of Eqs. (29) and (30), the coordinates of the FRF matrix of the coupled structure can be partitioned into three regions [7]:

1. Interior DOFs of substructure A, thus forward denoted with a subscript "a"
2. Interior DOFs of substructure B, thus forward denoted with a subscript "b"
3. The coupling DOFs of both substructures, thus forward denoted with a subscript "c".

With these three specific regions defined, the FRF for the complete system can be written as:

$$\begin{Bmatrix} \boldsymbol{X}_a \\ \boldsymbol{X}_c \\ \boldsymbol{X}_b \end{Bmatrix} = \begin{bmatrix} \boldsymbol{Y}_{aa} & \boldsymbol{Y}_{ac} & \boldsymbol{Y}_{ab} \\ \boldsymbol{Y}_{ca} & \boldsymbol{Y}_{cc} & \boldsymbol{Y}_{cb} \\ \boldsymbol{Y}_{ba} & \boldsymbol{Y}_{bc} & \boldsymbol{Y}_{bb} \end{bmatrix} \begin{Bmatrix} \boldsymbol{F}_a \\ \boldsymbol{F}_c \\ \boldsymbol{F}_b \end{Bmatrix} \tag{31}$$

Where the $\boldsymbol{X}$'s and $\boldsymbol{F}$'s are column vectors and the receptance block matrix will be called $\boldsymbol{Y}^C$. Simplifying the notation and expressing compatibility and equilibrium requirements yields the following:

$$\boldsymbol{X}_a = \boldsymbol{x}_i^A \tag{32}$$

$$\boldsymbol{X}_b = \boldsymbol{x}_i^A \tag{33}$$

$$\boldsymbol{X}_c = \boldsymbol{x}_c^A = \boldsymbol{x}_c^B \tag{34}$$

$$\boldsymbol{F}_c = \boldsymbol{f}_c^A + \boldsymbol{f}_c^B \tag{35}$$

$$\boldsymbol{F}_a = \boldsymbol{f}_i^A \tag{36}$$

$$\boldsymbol{F}_b = \boldsymbol{f}_i^B \tag{37}$$

Where the compatibility and equilibrium conditions are represented by Eqs. (34) and (35), respectively. Going back to Eqs. (29) and (30), the $\boldsymbol{x}_c^A$ and $\boldsymbol{x}_c^B$ terms can be expanded to find:

$$\boldsymbol{x}_c^A = \boldsymbol{Y}_{ci}^A \boldsymbol{f}_i^A + \boldsymbol{Y}_{cc}^A \boldsymbol{f}_c^A \tag{38}$$

$$\boldsymbol{x}_c^B = \boldsymbol{Y}_{ci}^B \boldsymbol{f}_i^B + \boldsymbol{Y}_{cc}^B \boldsymbol{f}_c^B \tag{39}$$

Then, the compatibility condition of Eq. (34) is used to arrive at a relationship between $\boldsymbol{f}_c^A$ and $\boldsymbol{f}_c^B$:

$$\boldsymbol{f}_c^B = \boldsymbol{Y}_{cc}^{B\,-1}\left(\boldsymbol{Y}_{cc}^A \boldsymbol{f}_c^A + \boldsymbol{Y}_{ci}^A \boldsymbol{f}_i^A - \boldsymbol{Y}_{ci}^B \boldsymbol{f}_i^B\right) \tag{40}$$

By utilizing the equilibrium condition of Eq. (35), along with Eqs. (36-39), the compatibility condition can be written as:

$$\boldsymbol{x}_c^A = \boldsymbol{Y}_{ci}^A \boldsymbol{F}_a + \boldsymbol{Y}_{cc}^A \boldsymbol{f}_c^A \tag{41}$$

$$\boldsymbol{x}_c^A = \boldsymbol{X}_c = \boldsymbol{Y}_{ca}\boldsymbol{F}_a + \boldsymbol{Y}_{cc}\left(\boldsymbol{f}_c^A + \boldsymbol{f}_c^B\right) + \boldsymbol{Y}_{cb}\boldsymbol{F}_b \tag{42}$$

Substituting Eq. (40) into (41) and then setting Eqs. (41) and (42) equal to each other produces the following after some simplification:

$$\boldsymbol{Y}_{ci}^A \boldsymbol{F}_a + \boldsymbol{Y}_{cc}^A \boldsymbol{f}_c^A = \dots = \left(\boldsymbol{Y}_{ca} + \boldsymbol{Y}_{cc}\boldsymbol{Y}_{cc}^{B\,-1}\boldsymbol{Y}_{ci}^A\right)\boldsymbol{F}_a + \left(\boldsymbol{Y}_{cc} + \boldsymbol{Y}_{cc}\boldsymbol{Y}_{cc}^{B\,-1}\boldsymbol{Y}_{cc}^A\right)\boldsymbol{f}_c^A + \left(\boldsymbol{Y}_{cb} - \boldsymbol{Y}_{cc}\boldsymbol{Y}_{cc}^{B\,-1}\boldsymbol{Y}_{ci}^B\right)\boldsymbol{F}_B \tag{43}$$

The corresponding force terms from each side of Eq. (43) are then set equal to each other:

$$\boldsymbol{Y}_{cc} + \boldsymbol{Y}_{cc}\boldsymbol{Y}_{cc}^{B\,-1}\boldsymbol{Y}_{cc}^A = \boldsymbol{Y}_{cc}^A \tag{44}$$

$$\boldsymbol{Y}_{ca} + \boldsymbol{Y}_{cc}\boldsymbol{Y}_{cc}^{B\,-1}\boldsymbol{Y}_{ci}^A = \boldsymbol{Y}_{ci}^A \tag{45}$$

$$\boldsymbol{Y}_{cb} - \boldsymbol{Y}_{cc}\boldsymbol{Y}_{cc}^{B\,-1}\boldsymbol{Y}_{ci}^B = \boldsymbol{0} \tag{46}$$

Equations (44–46) can be solved for the $\boldsymbol{Y}_{cc}$ term in the following steps:

$$\boldsymbol{Y}_{cc}\left(\boldsymbol{I} + \boldsymbol{Y}_{cc}^{B\,-1}\boldsymbol{Y}_{cc}^A\right) = \boldsymbol{Y}_{cc}^A$$

$$\boldsymbol{Y}_{cc} = \boldsymbol{Y}_{cc}^A\left(\boldsymbol{I} + \boldsymbol{Y}_{cc}^{B\,-1}\boldsymbol{Y}_{cc}^A\right)^{-1}$$

$$\boldsymbol{Y}_{cc} = \boldsymbol{Y}_{cc}^A\left(\boldsymbol{Y}_{cc}^{B\,-1}\left(\boldsymbol{Y}_{cc}^B\boldsymbol{I} + \boldsymbol{Y}_{cc}^B\boldsymbol{Y}_{cc}^{B\,-1}\boldsymbol{Y}_{cc}^A\right)\right)^{-1}$$

$$\boldsymbol{Y}_{cc} = \boldsymbol{Y}_{cc}^A\left(\boldsymbol{Y}_{cc}^B + \boldsymbol{Y}_{cc}^A\right)^{-1}\boldsymbol{Y}_{cc}^B = \boldsymbol{Y}_{cc}^B\left(\boldsymbol{Y}_{cc}^B + \boldsymbol{Y}_{cc}^A\right)^{-1}\boldsymbol{Y}_{cc}^A$$

Note that the terms of the final expression for $Y_{cc}$ can be rearranged in either order since the matrices are all symmetric in theory. Then, through the steps described for the simplified coupling problem, the final expression for $Y_{cc}$ can be put in the form of Eq. (26):

$$Y_{cc} = \mathbf{Y}^{A}_{cc} + \mathbf{Y}^{A}_{cc}\left(\mathbf{Y}^{B}_{cc} + \mathbf{Y}^{A}_{cc}\right)^{-1}\mathbf{Y}^{A}_{cc} \tag{47}$$

Equation (47) is then used to solve Eqs. (45) and (46) for $Y_{ca}$ and $Y_{cb}$ respectively. The remainder of the entries of the $Y^{C}$ receptance block matrix are found through similar applications of eqs. (31–36). It is worth noting that $Y_{ca}$ is the transpose of $Y_{ac}$, $Y_{ba}$ is the transpose of $Y_{ab}$, and so on for the other entries of $Y^{C}$.

The remainder of the receptance terms for $Y^{C}$ are shown in Eqs. (48–53):

$$Y_{aa} = \mathbf{Y}^{A}_{ii} - \mathbf{Y}^{A}_{ic}\left(\mathbf{Y}^{A}_{cc} + \mathbf{Y}^{B}_{cc}\right)^{-1}\mathbf{Y}^{A}_{ci} \tag{48}$$

$$Y_{ab} = \mathbf{Y}^{A}_{ic}\left(\mathbf{Y}^{A}_{cc} + \mathbf{Y}^{B}_{cc}\right)^{-1}\mathbf{Y}^{B}_{ci} \tag{49}$$

$$Y_{bb} = \mathbf{Y}^{B}_{ii} - \mathbf{Y}^{B}_{ic}\left(\mathbf{Y}^{A}_{cc} + \mathbf{Y}^{B}_{cc}\right)^{-1}\mathbf{Y}^{B}_{ci} \tag{50}$$

$$Y_{ba} = \mathbf{Y}^{B}_{ic}\left(\mathbf{Y}^{A}_{cc} + \mathbf{Y}^{B}_{cc}\right)^{-1}\mathbf{Y}^{A}_{ci} \tag{51}$$

$$Y_{ca} = \left[Y_{ac}\right]^{T} = \mathbf{Y}^{A}_{ci} - \mathbf{Y}^{A}_{cc}\left(\mathbf{Y}^{A}_{cc} + \mathbf{Y}^{B}_{cc}\right)^{-1}\mathbf{Y}^{A}_{ci} \tag{52}$$

$$Y_{cb} = \left[Y_{bc}\right]^{T} = \mathbf{Y}^{A}_{cc}\left(\mathbf{Y}^{A}_{cc} + \mathbf{Y}^{B}_{cc}\right)^{-1}\mathbf{Y}^{B}_{ci} \tag{53}$$

Equations (47–53) can then be reassembled into the canonical form of the FBS method as developed in [2] and shown in Eq. (53).

$$\mathbf{Y}^{C} = \begin{bmatrix} Y_{aa} & Y_{ac} & Y_{ab} \\ Y_{ca} & Y_{cc} & Y_{cb} \\ Y_{ba} & Y_{bc} & Y_{bb} \end{bmatrix} = \begin{bmatrix} \mathbf{Y}^{A}_{ii} & \mathbf{Y}^{A}_{ic} & \mathbf{0} \\ \mathbf{Y}^{A}_{ci} & \mathbf{Y}^{A}_{cc} & \mathbf{0} \\ \mathbf{0} & \mathbf{0} & \mathbf{Y}^{B}_{ii} \end{bmatrix} - \begin{bmatrix} \mathbf{Y}^{A}_{ic} \\ \mathbf{Y}^{A}_{cc} \\ -\mathbf{Y}^{B}_{ic} \end{bmatrix} \left(\mathbf{Y}^{A}_{cc} + \mathbf{Y}^{B}_{cc}\right)^{-1} \begin{bmatrix} \mathbf{Y}^{A}_{ic} \\ \mathbf{Y}^{A}_{cc} \\ -\mathbf{Y}^{B}_{ic} \end{bmatrix}^{T} \tag{54}$$

Equation (54) is the form of the FBS method that most commonly appears throughout literature and has served as the basis for many improvements to the method.

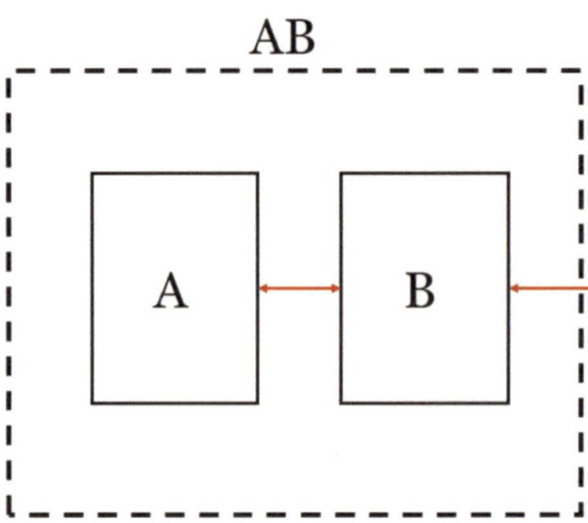


**Figure 3.**
*Three substructures to be coupled together. First, substructures A and B are coupled. Then, the combined substructure AB is couple with C.*

### 3.4 Application of FBS to multiple substructures

Though the FBS method as presented here is for the coupling of two substructures, the concept can be extended to include an arbitrarily large number of substructures. As presented in [8], the problem of coupling multiple substructures through the FBS method can be described by a simple example consisting of only three substructures. **Figure 3** demonstrates the general concept.

As shown in **Figure 3**, substructures A and B are first coupled to each other and then the combined substructure, AB, is coupled to C. In this way, the FBS method can be extend to an arbitrarily large number of substructures [8]. This same concept can also be applied to decoupling as shown in [9]. However, if substructure C was also coupled to A, or if all three substructures shared a coupling point, the process would no longer be as straightforward as presented here.

### 3.5 Generalized receptance coupling

A more generalized formulation of FBS was presented in [10]. The method was termed the "generalized receptance method" or "GRC" and was a physically and mathematically generalized version of the classical FBS/receptance method. In fact, the classical FBS method of [2] is a special case of GRC [11, 12]. However, most future work focused on either improvements to the classical FBS method, or entirely different coupling formulations such as in [9]. Thus, the mathematical formulation of GRC will not be presented here.

### 3.6 Lagrange multiplier frequency based substructuring (LM FBS)

In [9] the dual formulation of the base FBS method is improved through the inclusion of Lagrange multipliers. The dual assembly in the physical domain was introduced in a previous section, where a Lagrange multiplier term was introduced to enforce the equilibrium condition. The following analysis similarly uses Lagrange multipliers in the frequency domain.

With the inclusion of the Lagrange multipliers, the local equilibrium equations can be written in the frequency domain as:

$$\boldsymbol{ZX} + \boldsymbol{B}^T\boldsymbol{\lambda} = \boldsymbol{F} \tag{55}$$

Recall that $\boldsymbol{Z}$ is the dynamic stiffness (impedance could be used alternatively), $\boldsymbol{X}$ is the vector of displacements, $\boldsymbol{B}$ is the Boolean mapping matrix as previously defined, $\boldsymbol{\lambda}$ is a vector of Lagrange multipliers, and $\boldsymbol{F}$ is a vector containing the internal and interface forces.

The compatibility condition is expressed as:

$$\boldsymbol{BU} = \boldsymbol{0} \tag{56}$$

Where the Boolean matrix is constructed in theory to satisfy this equation. The transpose of this Boolean mapping matrix is used in Eq. (55) to define on which DOFs the connecting forces, $\boldsymbol{\lambda}$, will be applied.

The behavior of the complete system in the dual assembly formulation is described, as follows, by combining Eqs. (55) and (56):

$$\begin{bmatrix} Z & B^T \\ B & 0 \end{bmatrix} \begin{Bmatrix} U \\ \lambda \end{Bmatrix} = \begin{Bmatrix} F \\ 0 \end{Bmatrix} \tag{57}$$

Solving the first row of Eq. (57), the equilibrium equations, for $\boldsymbol{X}$ yields:

$$U = Y(F - B^T\lambda) \tag{58}$$

Where, as before, $\boldsymbol{Y}$ is the receptance (or admittance) and is the inverse of the dynamic stiffness, $\boldsymbol{Z}$. Eq. (58) can then be substituted into the second line of Eq. (57), the compatibility equations, to obtain an expression for $\boldsymbol{\lambda}$:

$$\lambda = \left(BYB^T\right)^{-1}BYF \tag{59}$$

Equation (59) can then be substituted back into Eq. (58) to obtain an expression for $\boldsymbol{X}$ independent of $\boldsymbol{\lambda}$:

$$U = Y\left(I - B^T\left(BYB^T\right)^{-1}BY\right)F \tag{60}$$

This formulation is mathematically equivalent to the classical FBS formulation. A full discussion and comparison of the two methods to demonstrate the equivalence is shown in [9]. Essentially, the only difference between the two methods is the point during the analysis in which the interface degrees of freedom are no longer considered for both coupled substructures. The choice is predetermined in the classical FBS method, whereas the LM FBS method allows the user to make the choice later in the analysis. The classical method eliminates the interface DOFs when not all information about the assembled system is known.

The ability to choose on which substructure the interface DOFs are retained provides the primary advantage of the LM FBS method over the classical FBS method. The user can then choose whichever set of FRFs contain a lower or less significant level of noise and, therefore, improve the quality of the coupled FRF.

With these FBS coupling methods described, the discussion will now turn to substructure decoupling in the frequency domain.

## 4. Decoupling theory and applications

Decoupling theory arose out of the need to describe the dynamics of a system or subsystem which may be difficult or impossible to measure through normal means. Examples of these types of systems are those that are very small, delicate, rotating, or potentially hazardous [13]. Some of the difficulties associated with these types of systems can be addressed through the use of non-contact testing methods such as laser Doppler vibrometry. However, non-contact methods come with their own difficulties and may still not be appropriate for the types of systems mentioned previously. Thus, decoupling theory and experimental application has still been an area of interest in recent years.

Substructure decoupling techniques can be generally classified as either "*inverse coupling*" techniques or "*direct decoupling techniques*" [14, 15]. Inverse coupling works by rearranging the equations for coupling so that one of the substructures is isolated and treated as unknown instead of the coupled structure. Most commonly, the FBS (or impedance) methods are used in inverse coupling [13, 16, 17].

On the other hand, direct decoupling uses a "fictitious substructure" which is added to the assembled system. This fictitious substructure is exactly the negative of the residual substructure which is defined as the substructure of the overall assembly which can be measured independently. This procedure starts from what is known as the 3-field formulation, so named for the sets of equations describing the system: one set to express the dynamic balance of the assembled system and of the fictitious substructure separately, one set to enforce compatibility at the interface DOFs, and one set to enforce equilibrium at the interface DOFs [15].

Either a primal or a dual approach can be used to carry out direct decoupling (primal and dual approach concepts are outlined in the mathematical background section). Compatibility and equilibrium conditions can be required either at only the coupling DOFs, or at coupling DOFs and additional internal DOFs of the residual substructure. A "standard interface" is defined to be when the conditions are enforced at only the coupling DOFs, an "extended interface" to be when all coupling and some internal DOFs are included, and a "mixed interface" to only include some coupling DOFs and some internal DOFs [14, 15].

The following subsections present the standard methods for both inverse coupling and direct decoupling.

### 4.1 Inverse coupling - Re-structuring of FBS method for decoupling

With the background math presented in the preceding sections, the most apparent method for decoupling structures is carried out through a restructuring of the FBS formulation. Recall the final formulation of the FBS methodology shown in Eq. (54). The most upper left corner of the $\mathbf{Y}^C$ receptance matrix is also described by Eq. (48). According to [13], this equation can be reorganized to yield:

$$\boldsymbol{Y}_{cc}^{B} = \boldsymbol{Y}_{ci}^{A}\left(\boldsymbol{Y}_{ii}^{A} - \boldsymbol{Y}_{aa}\right)^{-1}\boldsymbol{Y}_{ic}^{A} - \boldsymbol{Y}_{cc}^{A} \tag{61}$$

Equation (61) shows that the FRFs of the uncoupled structure B can be determined by measurements made on structure A while the two structures are coupled. The equation also implies that for the FRFs of B to be determined, the relationships must be known between the remote excitation DOFs and the coupling DOFs for the structure A when it is not coupled to B. These "relationships" are described by the terms with superscript "A" in Eq. (61).

Through similar analysis, the lower most left corner of the $\mathbf{Y}^C$ receptance matrix (also described by Eq. (50)) can be rearranged as:

$$\boldsymbol{Y}_{ic}^{B} = \boldsymbol{Y}_{ba}\left[\boldsymbol{Y}_{ci}^{A}\right]^{-1}\left(\boldsymbol{Y}_{cc}^{B} - \boldsymbol{Y}_{cc}^{A}\right) \tag{62}$$

Where the $\boldsymbol{Y}_{cc}^{B}$ term can be substituted in from Eq. (61). This analysis shows that the uncoupled FRFs of the structure B at DOFs remote from the coupling location can be found without direct measurement.

It is important to note that neither the auto FRF matrix ($\boldsymbol{Y}_{cc}^{A}$) nor the cross matrix ($\boldsymbol{Y}_{ba}$) need to be measured fully for Eqs. (61) and (62) respectively. The details of this reduction in necessary measurements effectively makes the measurement phase of any experiment less demanding and are discussed more fully in [13, 18].

Although Eqs. (61) and (62) are for the decoupling of structures, they were derived from the FBS (or impedance) method formulation. Thus, they are subject to the same issues encountered during the coupling procedures. The most common difficulties of decoupling using the previously described equations are listed briefly as follows and in more detail in [13]:

1. The coupling matrices which are to be inverted become singular due to ill conditioning of the data (i.e. signal noise and stiffness/damping properties at the measurement site)

2. Large order matrices are more susceptible to ill conditioning. With 6 DOFs needed at each connection point to have a full FRF matrix, a relatively small number of connection points can result in a large order matrix.

3. Modal truncation may occur if FRFs are synthesized from a modal database. This is often referred to as a "residual" problem, due to the residual dynamics truncated in the modal data. Like the rotational DOFs, this is an area in which further work has been done [19–21]

4. If any rotational DOFs are active in the coupling/decoupling, they need to be included in the FRF matrix. At the time of the publication of [13], measurement of rotational DOFs was still notoriously difficult. However, more work on the subject has been done in recent years and continues to be an area of improvement [20, 22–30].

Despite these difficulties, the decoupling method of Eqs. (61) and (62) does have some advantages over its coupling counterpart. These advantages primarily come from the ability to control certain aspects of the test through the design of the test fixture. Some of the advantages are control over the number of connection points, flexibility properties of one of the structures, and the location of the measurement points on the testing fixture [13].

In [31] a similar approach is explored, using impedance matrices instead of receptance. Furthermore, [31] shows that the decoupling equation is the same whether it is derived using impedances or mobilities, so long as only the coupling DOFs are used. However, if additional internal DOFs are included as well, the derivation of the decoupling equations is different depending on if impedance or mobility matrices are used.

### 4.2 Direct decoupling

As with previous mathematical descriptions in this chapter, it is assumed that two substructures, A and B, are coupled to form the assembled system, C. For the following descriptions of direct decoupling using the primal formulation, B as the residual substructure. It is important to note that the substructure B can be an individual substructure or an assembly of other substructures. With B as the known substructure, the goal is to extract the FRF of A from the coupled system, C, by cancelling out the dynamic effects of the residual substructure.

This cancelling out of the residual substructure is accomplished by adding a fictitious substructure to the coupled system. The fictitious substructure must have a dynamic stiffness equal to the negative of the residual substructure and must satisfy the compatibility and equilibrium conditions. The combined system of the coupled structure and the fictitious substructure [15] can be expressed in block diagonal format as:

$$\begin{bmatrix} \boldsymbol{Z}^C & \boldsymbol{0} \\ \boldsymbol{0} & -\boldsymbol{Z}^B \end{bmatrix} \begin{Bmatrix} \boldsymbol{x}^C \\ \boldsymbol{x}^B \end{Bmatrix} = \begin{Bmatrix} \boldsymbol{f}^C \\ \boldsymbol{f}^B \end{Bmatrix} + \begin{Bmatrix} \boldsymbol{g}^C \\ \boldsymbol{g}^B \end{Bmatrix} \tag{63}$$

Where, as previously, $\boldsymbol{f}$ corresponds to the external forces, $\boldsymbol{g}$ corresponds to the forces at the interfaces, $\boldsymbol{Z}$ is the dynamic stiffness, $\boldsymbol{x}$ is displacement, and the superscripts denote the corresponding structure.

This formulation implies that the interface between the fictitious substructure and the coupled system includes all of the internal DOFs of the residual substructure in addition to the coupling DOFs between the original substructures, A and B. However, it is not necessary to consider all of these DOFs to carry out the process. It is possible to decouple the system by considering only the coupling DOFs. Thus, the number of DOFs used when adding the fictitious substructure just has to be greater than or equal to the number of coupling DOFs between A and B. According to [15], there exist 3 options for these interface DOFs:

1. The "standard interface" which includes only the coupling DOFs between A and B

2. The "extended interface" which is the standard interface with the addition of a subset of internal DOFs from the residual subsystem

3. The "mixed interface" which includes a subset of the original coupling DOFs and a subset of the internal DOFs from the residual substructure.

Each interface has advantages and disadvantages depending upon the situation in which they are applied. For instance, as discussed later in the section on rotational DOFs, the mixed interface can be used to avoid including any rotational DOFs for which data is not available.

Regardless of which interface is used, the compatibility condition implies that any pair of matching DOFs between the coupled system and the residual substructure must have the same displacement. In other words, if there are matching DOFs $x_k^C$ and $x_l^B$, where $x_k^C$ is DOF number "*k*" on the coupled system and $x_l^B$ is DOF "*l*" on the residual substructure, then:

$$x_k^C - x_l^B = 0 \tag{64}$$

The compatibility condition can be expressed more generally as:

$$\begin{bmatrix} \boldsymbol{B}_C^C & \boldsymbol{B}_C^B \end{bmatrix} \begin{Bmatrix} \boldsymbol{x}^C \\ \boldsymbol{x}^B \end{Bmatrix} = \boldsymbol{0} \tag{65}$$

Where $\boldsymbol{B_C}$ is a signed Boolean matrix (with the subscript "*C*" referring to the compatibility condition) partitioned into portions corresponding to the coupled system, C, and the residual substructure, B. Each row of $\boldsymbol{B_C}$ corresponds to a pair of matching DOFs.

The equilibrium condition implies that the connecting forces at any pair of matching DOFs must sum to zero. Thus, if $g_k^C$ and $g_l^B$ are interface forces at DOF "*k*" on C and DOF "*l*" on B, then:

$$g_k^C + g_l^B = 0 \tag{66}$$

Additionally, the interface force for any DOF which does not belong to the equilibrium interface must be equal to zero. In other words, if "*m*" is an arbitrary DOF on either C or B and is not part of the equilibrium interface, then:

$$g_m^C = 0 \; or \; g_m^B = 0$$

The equilibrium conditions can be more generally expressed as:

$$\begin{bmatrix} \boldsymbol{L}_E^C \\ \boldsymbol{L}_E^B \end{bmatrix}^T \begin{Bmatrix} \boldsymbol{g}^C \\ \boldsymbol{g}^B \end{Bmatrix} = \boldsymbol{0} \tag{67}$$

Where the matrix $\boldsymbol{L}_E = \begin{bmatrix} \boldsymbol{L}_E^C & \boldsymbol{L}_E^B \end{bmatrix}$ is a Boolean localization matrix which is partitioned into portions corresponding to the coupled system, C, and the residual substructure, B. The subscript "*E*" specifies that the Boolean matrix is for the equilibrium condition.

Equations (63–67) together make up the previously mentioned 3-field formulation. With the 3-field formulation to use as a starting point, the following subsections present the primal and dual formulations for direct decoupling.

### *4.2.1 Direct decoupling - Primal formulation*

The primal formulation is defined for direct decoupling in the same way as for coupling in the physical domain (see Eq. (6)). Though the two presentations are mathematically equivalent, the primal formulation is defined again in this section to show it in partitioned form.

The primal formulation is based upon the definition of a unique set of DOFs:

$$\begin{Bmatrix} \boldsymbol{x}^C \\ \boldsymbol{x}^B \end{Bmatrix} = \begin{bmatrix} \boldsymbol{L}_C^C \\ \boldsymbol{L}_C^B \end{bmatrix} \{\boldsymbol{q}\} \tag{68}$$

Where $\boldsymbol{q}$ is the unique set of DOFs and $\boldsymbol{L}_C$ is a localization matrix (similar to $\boldsymbol{L}_E$) with the subscript "C" denoting that this matrix will be used in the compatibility condition.

The compatibility is satisfied for any $\boldsymbol{q}$ since, by definition, $\boldsymbol{q}$ is a unique set of DOFs. This concept can be shown mathematically as:

$$\begin{bmatrix} \boldsymbol{B}_C^C & \boldsymbol{B}_C^B \end{bmatrix} \begin{Bmatrix} \boldsymbol{x}^C \\ \boldsymbol{x}^B \end{Bmatrix} = \begin{bmatrix} \boldsymbol{B}_C^C & \boldsymbol{B}_C^B \end{bmatrix} \begin{bmatrix} \boldsymbol{L}_C^C \\ \boldsymbol{L}_C^B \end{bmatrix} \{\boldsymbol{q}\} = \boldsymbol{0} \forall \boldsymbol{q} \tag{69}$$

Which implies:

$$\begin{bmatrix} \boldsymbol{B}_C^C & \boldsymbol{B}_C^B \end{bmatrix} \begin{bmatrix} \boldsymbol{L}_C^C \\ \boldsymbol{L}_C^B \end{bmatrix} = \boldsymbol{0} \; or \; \begin{bmatrix} \boldsymbol{L}_C^C \\ \boldsymbol{L}_C^B \end{bmatrix}^T \begin{bmatrix} \boldsymbol{B}_C^C & \boldsymbol{B}_C^B \end{bmatrix}^T = \boldsymbol{0} \tag{70}$$

Equation (70) shows that $L_C$ is the null space of $B_C$ and, equivalently, $B_C{}^T$ is the null space of $L_C{}^T$.

With the compatibility condition (Eq. (65)) satisfied, the 3-field formulation reduces to:

$$\begin{cases} \begin{bmatrix} Z^C & 0 \\ 0 & -Z^B \end{bmatrix} \begin{bmatrix} L_C^C \\ L_C^B \end{bmatrix} \{q\} = \begin{Bmatrix} f^C \\ f^B \end{Bmatrix} + \begin{Bmatrix} g^C \\ g^B \end{Bmatrix} \\ \begin{bmatrix} L_E^C \\ L_E^B \end{bmatrix}^T \begin{Bmatrix} g^C \\ g^B \end{Bmatrix} = 0 \end{cases} \tag{71}$$

Following the derivation of [15], Eq. (71) is pre-multiplied by $L_E{}^T$ and Eq. (67) is used to reduce the formulation to:

$$\begin{bmatrix} L_E^C \\ L_E^B \end{bmatrix}^T \begin{bmatrix} Z^C & 0 \\ 0 & -Z^B \end{bmatrix} \begin{bmatrix} L_C^C \\ L_C^B \end{bmatrix} \{q\} = \begin{bmatrix} L_E^C \\ L_E^B \end{bmatrix}^T \begin{Bmatrix} f^C \\ f^B \end{Bmatrix} \tag{72}$$

Which rearranges to yield:

$$\{q\} = \left( \begin{bmatrix} L_E^C \\ L_E^B \end{bmatrix}^T \begin{bmatrix} Z^C & 0 \\ 0 & -Z^B \end{bmatrix} \begin{bmatrix} L_C^C \\ L_C^B \end{bmatrix} \right)^+ \begin{bmatrix} L_E^C \\ L_E^B \end{bmatrix}^T \begin{Bmatrix} f^C \\ f^B \end{Bmatrix} \tag{73}$$

Where the superscript "+" refers to what is called the "generalized inverse," or the "pseudoinverse." The pseudoinverse is an "inverse-like" operation that, unlike a standard inversion, can be used for an over determined matrix. In this case, a Moore-Penrose pseudoinverse is used.

According to [15], in order to obtain a determined or over-determined matrix for the primal formulation:

$$N_C \geq N_E > n_c \tag{74}$$

Where $N_C$ is the number of compatibility interface DOFs, $N_E$ is the number of equilibrium interface DOFs, and $n_c$ is the number of DOFs coupled between substructures A and B. If $N_C > N_E$, then Eq. (72) is satisfied by $\boldsymbol{q}$ only in a least-square sense (i.e. it is not satisfied exactly). However, due to the nature of the primal formulation, the compatibility condition *is* satisfied exactly by $\boldsymbol{q}$.

Finally, the FRF for the unknown substructure, A, can be found from Eq. (73) as:

$$Y^A = \left( \begin{bmatrix} L_E^C \\ L_E^B \end{bmatrix}^T \begin{bmatrix} Z^C & 0 \\ 0 & -Z^B \end{bmatrix} \begin{bmatrix} L_C^C \\ L_C^B \end{bmatrix} \right)^+ \begin{bmatrix} L_E^C \\ L_E^B \end{bmatrix}^T \tag{75}$$

$$Y^A = \left( \left[L_E^C\right]^T Z^C L_C^C - \left[L_E^B\right]^T Z^B L_C^B \right)^+ \left[ \left[L_E^C\right]^T \quad \left[L_E^B\right]^T \right] \tag{76}$$

Equations (75) and (76) conclude the derivation of the primal formulation for direct decoupling. Additionally, if the extended interface is used, $Y^A$ will contain some rows and columns which correspond to the internal DOFs of the residual substructure, B. These entries from the residual substructure are meaningless and should thus be discarded along with the duplicate entries for the interface DOFs during subsequent analysis. There are other derivations for decoupling via the primal formulations, such as that described in [32]. However, the differences for most of these formulations are minor and so they will not be presented independently here.

#### *4.2.2 Direct decoupling - Dual formulation*

In contrast to the primal formulation, the total set of DOFs is retained in the dual formulation. Thus, each interface DOFs is present in the formulation as many times as there are substructures connected at that interface point. The dual formulation also differs from the primal formulation in that the equilibrium condition is strictly enforced as opposed to the compatibility condition in the primal formulation. In other words, continuing with the same notation as in the previous section:

$$g_k^C + g_l^B = 0 \tag{77}$$

Which can be satisfied by the choice:

$$g_k^C = -\lambda \; and \; g_l^B = \lambda$$

Where $k$ and $l$ are arbitrary, coupled DOFs on the coupled system and residual substructure respectively. This enforcement of the equilibrium condition can be written more generally for the entire system as:

$$\begin{Bmatrix} \boldsymbol{g}^C \\ \boldsymbol{g}^B \end{Bmatrix} = -\begin{bmatrix} \left[\boldsymbol{B}_E^C\right]^T \\ \left[\boldsymbol{B}_E^B\right]^T \end{bmatrix} \{\lambda\} \tag{78}$$

Where $\lambda$ is a vector of Lagrange multipliers corresponding the interface force intensity and $\boldsymbol{B}_{\mathrm{E}}$ is a signed Boolean matrix which satisfies Eq. (78) by construction. With this formulation, the equilibrium condition of Eq. (67) can be rewritten as:

$$\begin{bmatrix} \boldsymbol{L}_E^C \\ \boldsymbol{L}_E^B \end{bmatrix}^T \begin{Bmatrix} \boldsymbol{g}^C \\ \boldsymbol{g}^B \end{Bmatrix} = -\begin{bmatrix} \boldsymbol{L}_E^C \\ \boldsymbol{L}_E^B \end{bmatrix}^T \begin{bmatrix} \left[\boldsymbol{B}_E^C\right]^T \\ \left[\boldsymbol{B}_E^B\right]^T \end{bmatrix} \{\lambda\} = \boldsymbol{0} \tag{79}$$

Similar to the primal formulation, $\boldsymbol{B}_E{}^T$ is the null space of $\boldsymbol{L}_E{}^T$, and $\boldsymbol{L}_E$ is the null space of $\boldsymbol{B}_E$. With the equilibrium condition (Eq. (67)) satisfied, the 3-field formulation reduces to:

$$\begin{cases} \begin{bmatrix} \boldsymbol{Z}^C & \boldsymbol{0} \\ \boldsymbol{0} & -\boldsymbol{Z}^B \end{bmatrix} \begin{Bmatrix} \boldsymbol{x}^C \\ \boldsymbol{x}^B \end{Bmatrix} + \begin{bmatrix} \left[\boldsymbol{B}_E^C\right]^T \\ \left[\boldsymbol{B}_E^B\right]^T \end{bmatrix} \{\lambda\} = \begin{Bmatrix} \boldsymbol{f}^C \\ \boldsymbol{f}^B \end{Bmatrix} \\ \begin{bmatrix} \boldsymbol{B}_C^C & \boldsymbol{B}_C^B \end{bmatrix} \begin{Bmatrix} \boldsymbol{x}^C \\ \boldsymbol{x}^B \end{Bmatrix} = \boldsymbol{0} \end{cases} \tag{80}$$

An expression for $\lambda$ can be found by solving Eq. (80) for the displacements, $\boldsymbol{x}$, and substituting into Eq. (65) [14]. The resulting expression is:

$$\{\lambda\} = \left( \begin{bmatrix} B_C^C & B_C^B \end{bmatrix} \begin{bmatrix} Z^C & 0 \\ 0 & -Z^B \end{bmatrix}^{-1} \begin{bmatrix} [B_E^C]^T \\ [B_E^B]^T \end{bmatrix} \right)^+ \begin{bmatrix} B_C^C & B_C^B \end{bmatrix} \begin{bmatrix} Z^C & 0 \\ 0 & -Z^B \end{bmatrix}^{-1} \begin{Bmatrix} f^C \\ f^B \end{Bmatrix} \tag{81}$$

Where again, the "+" superscript denotes the generalized or pseudo inverse.

As with the primal formulation, Eq. (74) must be satisfied to obtain a determined or over-determined matrix to carry out the generalized inverse. For the dual formulation if $N_C > N_E$, then the compatibility conditions (Eq. (65)) are only satisfied approximately. However, the equilibrium conditions are satisfied exactly by the construction of the dual formulation.

By substituting the expression for $\lambda$ (Eq. (81)) into Eq. (80), an expression for $x$ is obtained in the form:

$$\begin{Bmatrix} x^C \\ x^B \end{Bmatrix} = \begin{bmatrix} Z^C & 0 \\ 0 & -Z^B \end{bmatrix}^{-1} - \begin{bmatrix} Z^C & 0 \\ 0 & -Z^B \end{bmatrix}^{-1} \begin{bmatrix} [B_E^C]^T \\ [B_E^B]^T \end{bmatrix} \times \times \left( \begin{bmatrix} B_C^C & B_C^B \end{bmatrix} \begin{bmatrix} Z^C & 0 \\ 0 & -Z^B \end{bmatrix}^{-1} \begin{bmatrix} [B_E^C]^T \\ [B_E^B]^T \end{bmatrix} \right)^+$$
$$\begin{bmatrix} B_C^C & B_C^B \end{bmatrix} \begin{bmatrix} Z^C & 0 \\ 0 & -Z^B \end{bmatrix}^{-1} \begin{Bmatrix} f^C \\ f^B \end{Bmatrix} \tag{82}$$

Where the right-hand side of the equation has been split into three lines and the "×" symbols are used to denote matrix multiplication where one term will not fit on a single line. Recall that the receptance, $Y$, is the inverse of the dynamic stiffness and assume that Eq. (82) is of the form:

$$x = Yf \tag{83}$$

Thus, Eq. (82) provides the receptance FRF matrix for the unknown substructure, A:

$$Y^A = \begin{bmatrix} Y^C & 0 \\ 0 & -Y^B \end{bmatrix} - \begin{bmatrix} Y^C & 0 \\ 0 & -Y^B \end{bmatrix} \begin{bmatrix} [B_E^C]^T \\ [B_E^B]^T \end{bmatrix} \times \times \left( \begin{bmatrix} B_C^C & B_C^B \end{bmatrix} \begin{bmatrix} Y^C & 0 \\ 0 & -Y^B \end{bmatrix} \begin{bmatrix} [B_E^C]^T \\ [B_E^B]^T \end{bmatrix} \right)^+$$
$$\begin{bmatrix} B_C^C & B_C^B \end{bmatrix} \begin{bmatrix} Y^C & 0 \\ 0 & -Y^B \end{bmatrix} \tag{84}$$

Where, again, the right-hand side of the equation has been split into three lines. Note that Eq. (84), the dual formulation for direct decoupling, all rows and columns in $Y^A$ which correspond to the interface DOFs appear twice (once for the coupled system and once for the residual substructure). As with the primal formulation, if the extended interface is used, $Y^A$ will contain some rows and columns which correspond to the internal DOFs of the residual substructure, B. Again, these entries are meaningless and should thus be discarded along with the duplicate entries for the interface DOFs during subsequent analysis.

This concludes the derivation of the dual formulation for direct decoupling. An alternative approach to the derivation of the dual formulation was presented in [32]. Since the end result of the two approaches is the same, the alternative derivation will not be presented here, but can be found in [32] ('a dual approach"). This method for decoupling via the dual formulation is mathematically equivalent to the LM FBS method for dynamic substructure proposed by [9] (Eq. (82) compared to Eq. (83)).

#### *4.2.3 Comparison of direct decoupling through primal and dual formulations*

Discussion of primal and dual formulations for direct decoupling will be concluded with a comparison of the methods. The receptance of the unknown substructure can be rewritten from Eq. 47 in condensed form as:

$$\boldsymbol{Y}_P^A = \left(\boldsymbol{L}_E^T\,\boldsymbol{Z}\,\boldsymbol{L}_C\right)^{+}\boldsymbol{L}_E^T \tag{85}$$

Where the subscript "*P*" specifies that the receptance is according to the primal formulation. If $N_C = N_E$ (number of compatibility and equilibrium interface points are equal), then a standard inversion can be used in place of the pseudoinverse, and Eq. (85) can be rearranged to yield:

$$\left(\boldsymbol{L}_E^T\,\boldsymbol{Z}\,\boldsymbol{L}_C\right)\boldsymbol{Y}_P^A = \boldsymbol{L}_E^T \tag{86}$$

Similarly, the dual formulation of Eq. (84) can be displayed in condensed form as:

$$\boldsymbol{Y}_D^A = \boldsymbol{Y} - \boldsymbol{Y}\,\boldsymbol{B}_E^T\left(\boldsymbol{B}_C\boldsymbol{Y}\,\boldsymbol{B}_E^T\right)^{-1}\boldsymbol{B}_C\,\boldsymbol{Y} \tag{87}$$

Where the subscript "*D*" specifies accordance to the dual formulation. The pseudoinverse has been replaced again by the standard inverse which implies the assumption that $N_C = N_E$. Pre-multiplication of Eq. (87) by $\boldsymbol{Z}$ and $\boldsymbol{L}_E^{\mathrm{T}}$ (in order) yields:

$$\boldsymbol{L}_E^T\,\boldsymbol{Z}\,\boldsymbol{Y}_D^A = \boldsymbol{L}_E^T - \boldsymbol{L}_E^T\,\boldsymbol{B}_E^T\left(\boldsymbol{B}_C\boldsymbol{Y}\,\boldsymbol{B}_E^T\right)^{-1}\boldsymbol{B}_C\,\boldsymbol{Y} \tag{88}$$

Recall from the previous section that $\boldsymbol{B}_E^{\mathrm{T}}$ is the null space of $\boldsymbol{L}_E^{\mathrm{T}}$:

$$\boldsymbol{L}_E^T\boldsymbol{B}_E^T = \boldsymbol{0}$$

Thus Eq. (88) is simplified to:

$$\boldsymbol{L}_E^T\,\boldsymbol{Z}\,\boldsymbol{Y}_D^A = \boldsymbol{L}_E^T \tag{89}$$

A comparison of Eq. (86) and (89) shows that:

$$\boldsymbol{Y}_D^A = \boldsymbol{L}_C\boldsymbol{Y}_P^A \tag{90}$$

Equation (90) shows that if $N_C = N_E$, then the primal and dual formulations are equivalent (where $N_C$ is the number of compatibility DOFs and $N_E$ is the number of equilibrium DOFs). However, due to the definition of the pseudoinverse (shown in the primal formulation section), Eq. (89) cannot be obtained for an overdetermined system. Thus, if $N_C > N_E$, the primal and dual formulations will produce different results [15].

#### *4.2.4 Direct decoupling via the hybrid formulation*

More recently, another method for decoupling has been developed which can be thought of as a combination of the primal and dual formulations [33]. This combined method has been termed the "direct hybrid assembly" or the "hybrid formulation for decoupling." Like the primal and dual formulations, the hybrid formulation is a method of direct decoupling, where a fictitious substructure is added to the coupled system to cancel out the dynamic effects of the residual substructure.

As mentioned previously, if the extended interface is used with primal or dual formulations, the equilibrium or compatibility conditions, respectively, may only be satisfied in a least square sense. In contrast, the hybrid formulation satisfies both the compatibility and equilibrium conditions exactly, regardless of interface choice. The direct hybrid assembly begins the same as the primal formulation for direct decoupling:

$$\begin{Bmatrix} \boldsymbol{x}^C \\ \boldsymbol{x}^B \end{Bmatrix} = \begin{bmatrix} \boldsymbol{L}_C^C \\ \boldsymbol{L}_C^B \end{bmatrix} \{\boldsymbol{q}\} \tag{91}$$

$$\begin{bmatrix} \boldsymbol{B}_C^C & \boldsymbol{B}_C^B \end{bmatrix} \begin{Bmatrix} \boldsymbol{x}^C \\ \boldsymbol{x}^B \end{Bmatrix} = \begin{bmatrix} \boldsymbol{B}_C^C & \boldsymbol{B}_C^B \end{bmatrix} \begin{bmatrix} \boldsymbol{L}_C^C \\ \boldsymbol{L}_C^B \end{bmatrix} \{\boldsymbol{q}\} = \boldsymbol{0} \forall \boldsymbol{q} \tag{92}$$

Recall that $\boldsymbol{q}$ is a unique set of DOFs, $\boldsymbol{L}_C$ is a Boolean localization matrix, and $\boldsymbol{B}_C$ is a signed Boolean matrix defining the coupled DOFs. By definition, this choice guarantees the compatibility condition is satisfied exactly. It is then possible to also satisfy the equilibrium exactly by proceeding in the fashion of the dual formulation. The connecting forces are written as:

$$\begin{bmatrix} \boldsymbol{L}_C^C \\ \boldsymbol{L}_C^B \end{bmatrix} \{\boldsymbol{q}\} + \begin{bmatrix} \boldsymbol{Y}^C & \boldsymbol{0} \\ \boldsymbol{0} & -\boldsymbol{Y}^B \end{bmatrix} \begin{bmatrix} \left[\boldsymbol{B}_E^C\right]^T \\ \left[\boldsymbol{B}_E^B\right]^T \end{bmatrix} \{\boldsymbol{\lambda}\} = \begin{bmatrix} \boldsymbol{Y}^C & \boldsymbol{0} \\ \boldsymbol{0} & -\boldsymbol{Y}^B \end{bmatrix} \begin{Bmatrix} \boldsymbol{f}^C \\ \boldsymbol{f}^B \end{Bmatrix} \tag{93}$$

Which can be shown in condensed form as:

$$\boldsymbol{L}_C \boldsymbol{q} + \boldsymbol{Y} \boldsymbol{B}_E^T \{\boldsymbol{\lambda}\} = \boldsymbol{Y} \boldsymbol{f} \tag{94}$$

If $\boldsymbol{q}$ and $\boldsymbol{\lambda}$ are gathered into a single vector, the generalized inverse can be used to arrive at:

$$\begin{Bmatrix} \boldsymbol{q} \\ \boldsymbol{\lambda} \end{Bmatrix} = \begin{bmatrix} \boldsymbol{L}_C & \boldsymbol{Y} \boldsymbol{B}_E^T \end{bmatrix}^+ \boldsymbol{Y} \boldsymbol{f} \tag{95}$$

Note that for the generalized inverse operation, the matrix must be determined or overdetermined which requires $N_C \geq N_E \geq n_C$ (where $N_C$ is number of interface DOFs where compatibility is enforced, $N_E$ is the same for equilibrium respectively, and $n_C$ is the actual number of coupling DOFs). If $N_C > N_E$, Eq. 61 is only satisfied in a least-square sense, while compatibility and equilibrium are satisfied exactly.

Eq. (95) can be rewritten in a more physically intuitive form as:

$$\begin{Bmatrix} \boldsymbol{q} \\ \boldsymbol{\lambda} \end{Bmatrix} = \begin{bmatrix} \boldsymbol{Y}^A \\ \boldsymbol{T} \end{bmatrix} \boldsymbol{f} \tag{96}$$

Where $\boldsymbol{Y}^A$ is the FRF matrix of the unknown subsystem and $\boldsymbol{T}$ is the transmissibility between the external forces and the connecting force intensities, $\boldsymbol{\lambda}$.

In this hybrid formulation, columns of $\boldsymbol{Y}^A$ corresponding to equilibrium interface DOFs appear twice. Also, $\boldsymbol{Y}^A$ contains rows and columns corresponding to the residual substructure when an extended interface is used. These rows and columns are meaningless in the context of the FRF of the unknown substructure and so should be discarded during analysis along with redundant entries.

In [33], it is further shown that hybrid and dual formulations are mathematically equivalent if $N_C = N_E$. However, if $N_C \neq N_E$ the two formulations provide different results. Based on a simple case study performed in [33], in cases where the two formulations gave different results, there was no clear indication of one method producing better results than the other.

## 5. Conclusion

The derivations and references in this chapter provide the foundation upon which FBS coupling and decoupling techniques are built. These techniques allow for dynamic information from individual substructures to be combined to provide the dynamics of a built-up structure/system. Alternatively, other FBS techniques allow for the dynamics of one substructure to be obtained indirectly from those of the overall structure and the "residual" substructure. As computational and experimental capabilities improve, new and updated approaches for coupling and decoupling techniques continue to arise [19–30, 33–40]. In addition to the application of existing techniques in new fields, a few of the main theoretical issues which have been (and continue to be) addressed are: ill conditioning of matrices, spurious peaks in the coupled/decoupled FRFs, inclusion of rotational DOFs, and uncertainty propagation from measured to calculated FRFs. For any particular application of the techniques presented in this chapter, the next step should always be an investigation of the experimental difficulties encountered by others in similar applications, as well as any accompanying improvements which have been developed to account for said difficulties.

## Acknowledgements

Parts of this chapter were previously published in the doctoral thesis by the same author: Matthew C. Clontz. Genetic characterization of "*A Novel Method for Vibration Analysis of the Tire-Vehicle System via Frequency Based Substructuring.*" [2018]. [Unpublished doctoral thesis]. Virginia Tech.

Available from: https://vtechworks.lib.vt.edu/items/8676f319-63da-4d7f-bcac-9976ad46d0c3

## Appendices and nomenclature

| | |
|---|---|
| M | mass matrix |
| C | damping matrix |
| K | stiffness matrix |

| | |
|---|---|
| x | displacement vector |
| f | external forcing function |
| g | internal forcing function (between coupled DOFs) |
| B | signed boolean matrix which operates on the coupling DOFs |
| L | signed boolean matrix relating to either compatibility or equilibrium conditions |
| q | generalized set of DOFs |
| λ | lagrange multipliers corresponding to the equilibrium condition |
| Z | dynamic stiffness matrix |
| Y | receptance FRF matrix |
| $\boldsymbol{Y}_{ic}^{B}$ | receptance FRF matrix for substructure "B" between internal, "i," and coupled, "c," DOFs |
| $\boldsymbol{x}_{c}^{A}$ | displacement for coupled, "c," DOFs for substructure "A" |

## Author details

Matthew Clontz
Western Carolina University, Cullowhee, NC, USA

*Address all correspondence to: clontzm@wcu.edu

## References


[1] De Klerk D, Rixen DJ, Voormeeren S. General framework for dynamic substructuring: History, review and classification of techniques. AIAA Journal. 2008;**46**(5):1169-1181

[2] Jetmundsen B, Bielawa RL, Flannelly WG. Generalized frequency domain substructure synthesis. Journal of the American Helicopter Society. 1988;**33**(1):55-64

[3] Voormeeren S. Coupling procedure improvement & uncertainty quantification in experimental dynamic substructuring [Master's thesis]. TU Delft; 2007

[4] De Klerk D, Rixen D. Effektive hybride Analyse von gekoppelter mechanischer Subsysteme. Delft, Netherlands: TU Delft; [Master's Thesis]. 2004

[5] Farhat C, Hetmaniuk U. A fictitious domain decomposition method for the solution of partially axisymmetric acoustic scattering problems. Part I: Dirichlet boundary conditions. International Journal for Numerical Methods in Engineering. 2002;**54**(9): 1309-1332

[6] Ewins D. Modal Testing: Theory, Practice and Application. Vol. 171. Baldock, Hertfordshire, England: Research Studies Press LTD.; 2000. pp. 415-437

[7] Urgueira APV. Dynamic Analysis of Coupled Structures Using Experimental Data. London, UK: University of London; 1989

[8] Wang Z-W, Wang J. Inverse substructure method of three-substructures coupled system and its application in product-transport-system. Journal of Vibration and Control. 2010; **17**(6):943-951

[9] de Klerk D, Rixen DJ, de Jong J. The frequency based substructuring (FBS) method reformulated according to the dual domain decomposition method. In: 24th International Modal Analysis Conference. Vol. 36. New York: Springer; 2006

[10] Ren Y, Beards C. On substructure synthesis with FRF data. Journal of Sound and Vibration. 1995;**185**(5): 845-866

[11] Ren Y. The analysis and identification of friction joint parameters in the dynamic response of structures. [Ph.D. thesis]. UK: Imperial College of Science, Technology and Medicine, University of London; 1992

[12] Ren Y, Beards C. A generalized receptance coupling technique. In: Proceedings-SPIE the International Society for Optical Engineering. San Diego, California, United States: SPIE International Society for Optical; 1993. pp. 868-868

[13] Ind P, Ewins D. Impedance based decoupling and its application to indirect modal testing and component measurement: A numerical investigation. In: Proceedings of the Twenty First International Modal Analysis Conference, Kissimmee, FL. 2003

[14] D'Ambrogio W, Fregolent A. Decoupling procedures in the general framework of frequency based substructuring. In: Proceedings of 27th IMAC. Orlando (USA). 2009

[15] D'Ambrogio W, Fregolent A. Direct decoupling of substructures using primal and dual formulation. In: Linking Models and Experiments. Vol. 2. New York, New York, USA: Springer; 2011. pp. 47-76

[16] Sjövall P, Abrahamsson T. Substructure system identification from coupled system test data. Mechanical Systems and Signal Processing. 2008; **22**(1):15-33

[17] D'Ambrogio W, Fregolent A. Promises and pitfalls of decoupling procedures. In: Proceeding of 26th IMAC. Orlando (USA). 2008

[18] Ind P. The non-intrusive modal testing of delicate and critical structures. Imperial college of science, technology and medicine [PhD thesis]; 2004

[19] Nicgorski D, Avitabile P. Conditioning of FRF measurements for use with frequency based substructuring. Mechanical Systems and Signal Processing. 2010;**24**(2):340-351

[20] O'Callahan J, Avitabile P, Riemer R. System equivalent reduction expansion process (SEREP). In: Proceedings of the 7th International Modal Analysis Conference. Vol. 1. NY: Union College Schnectady; 1989. pp. 29-37

[21] Nicgorski D, Avitabile P. Experimental issues related to frequency response function measurements for frequency-based substructuring. Mechanical Systems and Signal Processing. 2010;**24**(5):1324-1337

[22] Balmes E, Billet L. Using expansion and interface reduction to enhance structural modification methods. In: IMAC-XIX: A Conference on Structural Dynamics. Vol. 1. Orlando, Florida: IMAC; 2001. pp. 615-621

[23] D'Amato E, Durante F, D'Ambrogio W. Distributed structural modification using local interface model. In: ISMA 2002-International Conference on Noise and Vibration Engineering. Leuven, Belgium: ISMA; 2002. pp. 1731-1739

[24] D'Ambrogio W, Sestieri A. A unified approach to substructuring and structural modification problems. Shock and Vibration. 2004;**11**(3–4):295-309

[25] de Klerk D, Rixen D, Voormeeren S, Pasteuning F. Solving the RDoF problem in experimental dynamic substructuring. In: 26th International Modal Analysis Conference (IMAC XXVI), Orlando, FL. 2008

[26] Pasteuning F. Experimental dynamic substructuring and its application to automotive research [Master's thesis]. TU Delft: Department of Engineering Mechanics; 2007

[27] De Klerk D. Dynamic Response Characterization of Complex Systems through Operational Identification and Dynamic Substructuring-an Application to Gear Noise Propagation in the Automotive Industry. TU Delft: Delft University of Technology; 2009

[28] Mayes RL. Tutorial on experimental dynamic substructuring using the transmission simulator method. In: Topics in Experimental Dynamics Substructuring and Wind Turbine Dynamics. Vol. 2. New York, New York, USA: Springer; 2012. pp. 1-9

[29] Mayes RL, Arviso M. Design studies for the transmission simulator method of experimental dynamic substructuring. In: International Seminar on Modal Analysis (ISMA2010). Leuven, Belgium: ISMA; 2010

[30] Mayes RL, Allen MS, Kammer DC. Correcting indefinite mass matrices due to substructure uncoupling. Journal of Sound and Vibration. 2013;**332**(22): 5856-5866

[31] D'Ambrogio W, Fregolent A. Prediction of substructure properties using decoupling procedures. In:

Proceedings of EURODYN. Paris, France: EURODYN; 2005

[32] Voormeeren S, De Klerk D, Rixen D. Uncertainty quantification in experimental frequency based substructuring. Mechanical Systems and Signal Processing. 2010;**24**(1):106-118

[33] D'Ambrogio W, Fregolent A. Are rotational DoFs essential in substructure decoupling? In: Dynamics of Coupled Structures. Vol. 1. Orlando, Florida, USA: Springer; 2014. pp. 27-36

[34] Chen F, Chen Y, Hua H. Vibration analysis of a submarine elastic propeller-shaft-hull system using FRF-based substructuring method. Journal of Sound and Vibration. 2019;**443**:460-482

[35] De Domenico D, Falsone G, Ricciardi G. Improved response-spectrum analysis of base-isolated buildings: A substructure-based response spectrum method. Engineering Structures. 2018;**162**:198-212

[36] Drozg A, Čepon G, Boltežar M. Full-degrees-of-freedom frequency based substructuring. Mechanical Systems and Signal Processing. 2018;**98**:570-579

[37] Hamedi B, Taheri S. Modular modeling of a half-vehicle system using generalized receptance coupling and frequency-based substructuring (GRCFBS). Vibration. 2024;**7**(4): 1063-1085

[38] Jalali MH, Rideout DG. Substructural damage detection using frequency response function based inverse dynamic substructuring. Mechanical Systems and Signal Processing. 2022;**163**: 108166

[39] Kodrič M, Čepon G, Boltežar M. Experimental framework for identifying inconsistent measurements in frequency-based substructuring. Mechanical Systems and Signal Processing. 2021;**154**:107562

[40] Mirza W, Kyprianou A, da Silva T, Rani M. Frequency based substructuring and coupling enhancement using estimated rotational frequency response functions. Experimental Techniques. 2024;**48**(3):423-437

**Chapter 5**

# Free Vibration Analysis of Functionally Graded Plates on Viscoelastic Foundations

*Abdelmoumen Anis Bousahla*

## Abstract

This chapter explores the free vibration behavior of temperature-dependent functionally graded plates (FGPs) resting on a viscoelastic foundation using a newly developed first-order shear deformation theory (FSDT). The proposed model features only four unknown variables and incorporates a parabolic distribution of transverse shear stress and strain through the plate's thickness, eliminating the need for a shear correction factor. Material properties graded across the plate's thickness are defined by a power-law distribution, and thermal effects are modeled using uniform, linear, non-linear, and sinusoidal temperature profiles applied to simply supported FGPs. Hamilton's principle and Navier's method are employed to derive motion equations and analytical solutions. This approach effectively predicts the frequencies of temperature-dependent FGPs, with numerical studies illustrating the influence of power-law indices, thermal fields, and damping coefficients on their dynamic behavior. The simplicity and accuracy of the proposed theory demonstrate its advancement over conventional FSDT models.

**Keywords:** free vibration, simple first-order shear, viscoelastic foundation, temperature-dependent properties, functionally graded plates

## 1. Introduction

This chapter builds upon research previously published in the journal *Structural Engineering and Mechanics* [1], where the free vibration behavior of temperature-dependent functionally graded plates (FGPs) resting on a viscoelastic foundation was investigated. That study employed a first-order shear deformation theory (FSDT) to analyze the effects of thermal fields, material gradation, and foundation parameters on the natural frequencies of FGPs. The findings highlighted the critical role of temperature-dependent properties in predicting the dynamic response of such structures. In this chapter, we extend that work by incorporating additional considerations related to material distribution and foundation-structure interactions, aiming to provide a more comprehensive understanding of the vibrational behavior of FGPs under complex environmental conditions.

Vibration is a fundamental phenomenon that affects the performance, reliability, and safety of structural components in a wide range of engineering applications. Understanding and controlling vibrations is particularly crucial in advanced structural systems exposed to dynamic and thermal environments. In recent decades, functionally graded materials (FGMs) have gained significant interest for vibration mitigation due to their ability to smoothly vary material properties across their thickness. Unlike conventional composites with abrupt interfaces between layers, FGMs exhibit a continuous gradation of properties between two or more constituent materials, typically metal and ceramic. This gradient design helps reduce thermal stress concentrations and improves the dynamic stability of structures by optimizing stiffness and damping characteristics [2, 3].

Given their unique advantages, FGMs are widely used in aerospace, automotive, civil engineering, and nuclear technologies, where dynamic loading and extreme thermal variations are common. Consequently, a substantial body of research has been dedicated to analyzing the vibrational behavior of functionally graded plates (FGPs), beams, and shells. The focus has been on predicting their dynamic responses, natural frequencies, and stability under different boundary conditions and loading scenarios, as these parameters are critical for ensuring structural integrity and longevity.

Various plate theories have been developed to model the vibrational responses of FGPs. Classical plate theory (CPT) is suitable for thin plates but neglects transverse shear deformation, limiting its accuracy in dynamic studies involving moderately thick or thick plates [4–11]. To overcome this limitation, first-order shear deformation theory (FSDT) incorporates shear deformation effects and provides better predictions for both thin and thick plates. However, FSDT assumes constant shear strain distribution, violating the equilibrium conditions at the top and bottom surfaces. Despite these simplifications, FSDT has been widely adopted due to its balance of accuracy and computational efficiency.

Several researchers have applied FSDT to investigate the free vibration and buckling behavior of FGPs. For example, Raju et al. [12] used the finite element method to analyze the free vibration of thin and moderately thick plates based on Mindlin's plate theory [13]. Liew et al. [14] extended this approach to study rectangular plates with various boundary conditions. Additional works, such as those by Della Croce and Venini [15] and Chen [16], examined bending and nonlinear vibrations under mechanical and thermal loads. The influence of elastic foundations on FGP vibration was explored by Fallah et al. [17], while Ganapathi et al. [18] combined FSDT with finite element modeling to investigate buckling.

In contrast to FSDT, higher-order shear deformation theories (HSDTs) offer improved accuracy by incorporating non-linear shear strain distributions. Researchers have developed simplified and refined versions of HSDTs to model the vibrational behavior of micro- and nano-structured FGPs [19–28]. Additionally, for thick plates, where the thickness stretching effect is significant, quasi-3D theories have been proposed to account for both normal and shear deformations [29, 30].

In recent years, considerable attention has been given to FGPs resting on elastic or viscoelastic foundations, as these conditions closely resemble real-world applications. Ebrahimi and Barati [31] investigated the hygrothermal effects on the vibration of viscoelastic FG nanobeams, highlighting the complex interplay between temperature, moisture, and dynamic response. Ghumare and Sayyad [32] introduced a fifth-order shear deformation theory to analyze bending under hygrothermal environments and elastic foundations. Nebab et al. [33] addressed the dynamic response using a four-variable plate theory, while Akbas et al. [34] studied FG porous beams subjected to pulse loads on viscoelastic supports. Despite these advances, there remains a need for simplified yet robust theories capable of accurately predicting the vibrational behavior of temperature-dependent FGPs on viscoelastic foundations.

This chapter aims to develop and apply a refined FSDT to investigate the free vibration response of FG plates with temperature-dependent material properties resting on a viscoelastic foundation. The proposed theory incorporates a parabolic distribution of transverse shear strain across the thickness, reducing computational complexity compared to HSDT and quasi-3D approaches. The temperature variation is assumed to occur along the thickness direction, and the material properties follow a power-law distribution. Employing Hamilton's principle, the governing equations of motion are derived, and the effects of thermal conditions, viscoelastic foundation parameters, and damping on the vibrational response are examined. This study provides insights relevant to the design of advanced aerospace structures subjected to dynamic and thermal challenges.

## 2. Theoretical developments

In this chapter, we consider a simply supported rectangular functionally graded plate (FGP) defined by its geometric parameters: length aaa, width bbb, and thickness hhh. The plate is analyzed using a Cartesian coordinate system (x,y,z) where x- and y-axes correspond to the longitudinal and transverse directions, respectively, while the z-axis represents the thickness coordinate. The mid-plane of the plate lies at z = 0, and the upper and lower surfaces are located at $z = \pm$, as illustrated in **Figure 1**.

The material composition of the plate is graded through the thickness according to a functionally graded material (FGM) distribution, typically modeled using a power-law or exponential gradation to describe the variation in properties from the top to the bottom surface. This gradation provides enhanced adaptability to different operational conditions by optimizing mechanical performance and resistance to thermal, mechanical, and hygrothermal effects.

The plate is assumed to rest on an elastic foundation characterized by the Winkler coefficient $k_w$ representing vertical stiffness, and the Pasternak coefficient $k_p$ accounting for shear interactions within the foundation. Additionally, the damping coefficient $c_t$ is considered to capture energy dissipation effects. These parameters significantly influence the mechanical and dynamic responses of the structure.

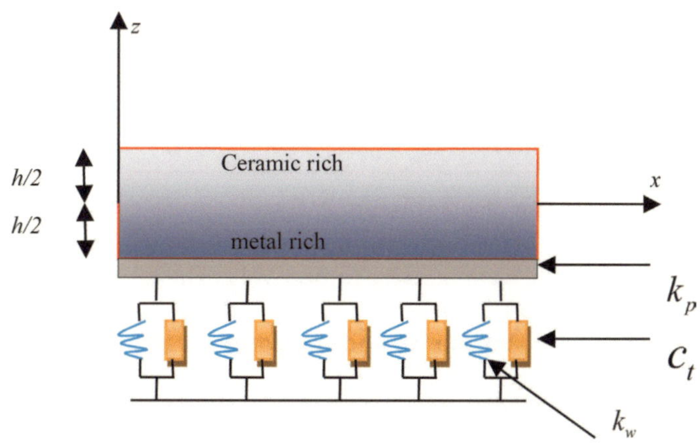


**Figure 1.**
*Material gradation of the FGP resting on a viscoelastic foundation.*

The simply supported boundary conditions considered in this study impose constraints on the displacement and rotation at the edges, aligning with classical plate theory or higher-order shear deformation models, depending on the complexity of the analysis. The established framework serves as a basis for comprehensive investigation of the plate's behavior under static loading, vibrational responses, and buckling instabilities, integrating the influence of nonlocal elasticity and various environmental interactions.

## 2.1 Displacement field and strain components

In the framework of the developed simplified first-order shear deformation theory (FSDT), the displacement field is constructed based on the following foundational assumptions:

1. The bending components of the in-plane displacements resemble those described by the Classical Plate Theory (CPT).

2. The transverse displacement is decomposed into bending and shear contributions to account for the effects of shear deformation through the plate thickness.

3. The in-plane displacement field consists of both bending and extensional components, providing a more comprehensive representation of the deformation behavior.

These assumptions allow for an improved prediction of mechanical responses, especially for moderately thick plates, where shear effects cannot be neglected. Unlike CPT, which assumes that normals to the mid-plane remain perpendicular after deformation, the FSDT incorporates shear deformation by relaxing this constraint, leading to more accurate modeling of deflections and stresses in functionally graded and composite materials.

In the current formulation, the general expression for the displacement field is represented as follows:

$$
\begin{aligned}
u(x,y,z,t) &= u_0(x,y,t) - z\frac{\partial w_b}{\partial x} \\
v(x,y,z,t) &= v_0(x,y,t) - z\frac{\partial w_b}{\partial y} \\
w(x,y,z,t) &= w_b(x,y,t) + w_s(x,y,t)
\end{aligned}
\tag{1}
$$

where $u$, $v$, and $w$ denote the displacements in the $x$, $y$, and $z$ directions, respectively; $u_0$, $v_0$, and $w_0$ are the mid-plane displacements; and $\theta_x$ and $\theta_y$ are the rotations of the transverse normal about the $y$ and $x$ axes, respectively. The inclusion of these rotational terms enriches the model, enhancing its applicability to advanced structural problems involving functionally graded and composite plates subjected to complex loading conditions.

This displacement formulation establishes the foundation for deriving strain–displacement relations and formulating the governing equations for bending, vibration, and stability analyses, while incorporating shear correction factors to address non-uniform shear distribution across the thickness.

$$\begin{Bmatrix} \varepsilon_x \\ \varepsilon_y \\ \gamma_{xy} \end{Bmatrix} = \begin{Bmatrix} \varepsilon_x^0 \\ \varepsilon_y^0 \\ \gamma_{xy}^0 \end{Bmatrix} + z \begin{Bmatrix} k_x^b \\ k_y^b \\ k_{xy}^b \end{Bmatrix}, \begin{Bmatrix} \gamma_{yz} \\ \gamma_{xz} \end{Bmatrix} = \begin{Bmatrix} \gamma_{yz}^0 \\ \gamma_{xz}^0 \end{Bmatrix} \tag{2}$$

where

$$\begin{Bmatrix} \varepsilon_x^0 \\ \varepsilon_y^0 \\ \gamma_{xy}^0 \end{Bmatrix} = \begin{Bmatrix} \dfrac{\partial u_0}{\partial x} \\ \dfrac{\partial v_0}{\partial x} \\ \dfrac{\partial u_0}{\partial y} + \dfrac{\partial v_0}{\partial x} \end{Bmatrix}, \begin{Bmatrix} k_x^b \\ k_y^b \\ k_{xy}^b \end{Bmatrix} = \begin{Bmatrix} -\dfrac{\partial^2 w_b}{\partial x^2} \\ -\dfrac{\partial^2 w_b}{\partial y^2} \\ -2\dfrac{\partial^2 w_b}{\partial x \partial y} \end{Bmatrix}, \begin{Bmatrix} \gamma_{yz}^0 \\ \gamma_{xz}^0 \end{Bmatrix} = \begin{Bmatrix} \dfrac{\partial w_s}{\partial y} \\ \dfrac{\partial w_s}{\partial x} \end{Bmatrix} \tag{3}$$

In the classical formulation of the First-Order Shear Deformation Theory (FSDT), the shear strain variation is assumed to be linear through the thickness of the plate. While this assumption simplifies the mathematical model, it deviates from the true physical behavior of shear stresses, which follow a parabolic distribution and naturally vanish at the top and bottom surfaces of the plate. Consequently, the conventional FSDT requires the application of a shear correction factor to compensate for the disparity between the assumed linear shear strain and the actual distribution of shear stresses.

To address this limitation, the present work introduces an enhanced shear deformation model by adopting a refined shear strain distribution function. This approach eliminates or significantly reduces the dependency on empirical correction factors, providing a more realistic and efficient description of shear deformation in functionally graded and composite plates. The proposed formulation captures the non-uniform shear behavior across the thickness while maintaining the simplicity and computational efficiency characteristic of FSDT. Based on these improvements, the shear strain vector is formulated as:

$$\begin{Bmatrix} \gamma^c{}_{yz} \\ \gamma^c{}_{xz} \end{Bmatrix} = g(z) \begin{Bmatrix} \gamma_{yz}^0 \\ \gamma_{xz}^0 \end{Bmatrix} \tag{4}$$

where $g(z)$ represents the function that describes the transverse shear strains distribution along the plate's thickness. This function is considered in order to attain stress-free boundary conditions on the plate's top and bottom surfaces. Also, the integration across the plate's thickness is similar to the shear correction factor used in the FSDT (5/6). The distribution of shear function can be taken as reported in [35]:

$$g(z) = \frac{5}{4} \cos\left(\frac{\pi z}{h}\right) \tag{5}$$

### 2.2 Constitutive relations

In this study, the variation of material properties in functionally graded plates (FGPs) is modeled using a power-law distribution of the constituent volume fractions. This distribution provides a smooth gradation of material properties from one surface of the plate to the other, offering enhanced flexibility in tailoring mechanical performance. The power-law model is particularly suitable for functionally graded materials (FGMs) where the volume fraction of the reinforcement and matrix phases varies continuously across the plate thickness.

The material properties, including elasticity modulus, thermal expansion coefficient, and thermal conductivity, are assumed to be temperature-dependent to account for the effects of thermal environments on structural performance. At any point within the plate, the effective material property are considered temperature-dependent, and positions can be defined as follows [21]:

$$\Gamma(z,T) = (\Gamma_c(T) - \Gamma_m(T))V_c + \Gamma_m(T) \text{ and } V_c(z) = \left(\frac{z}{h} + \frac{1}{2}\right)^p \tag{6}$$

where $\Gamma$ implies the FG plates' properties of the generic material, such as elastic modulus $E$, the Poisson's ratio $\nu$, coefficient of thermal expansion $\alpha$, and mass density $\rho$. Moreover, $C$ and $m$ are related to the pure ceramic and metal plates, respectively. $V_c$ denotes the ceramic volume fraction. The non-linear FG plate's material is given as reported as follows [36]:

$$P(T) = P_0\left(P_{-1}T^{-1} + 1 + P_1T + P_2T^2 + P_3T^3\right) \tag{7}$$

where $P(T)$ represents the temperature-dependent property of the material, and $T$ denotes the environmental temperature. The room temperature is considered as $T_0 = 300(K)$. The coefficients $P_{-1}$, $P_0$, $P_1$, $P_2$ and $P_3$ define the temperature-dependent variations of material properties unique to each constituent. The temperature variation $\Delta T(z)$ is assumed to occur exclusively across the thickness direction. The thermal conductivity $k$ is treated as temperature-independent. **Table 1** presents typical temperature-dependent properties for common functionally graded material components, including silicon nitride and stainless steel, as reported in [21].

The constitutive relations of an FGP are given as follows:

$$\begin{Bmatrix} \sigma_x \\ \sigma_y \\ \tau_{yz} \\ \tau_{xz} \\ \tau_{xy} \end{Bmatrix} = \begin{bmatrix} C_{11} & C_{12} & 0 & 0 & 0 \\ C_{12} & C_{22} & 0 & 0 & 0 \\ 0 & 0 & C_{44} & 0 & 0 \\ 0 & 0 & 0 & C_{55} & 0 \\ 0 & 0 & 0 & 0 & C_{66} \end{bmatrix} \begin{Bmatrix} \varepsilon_x \\ \varepsilon_y \\ \gamma_{yz} \\ \gamma_{xz} \\ \gamma_{xy} \end{Bmatrix} \tag{8}$$

where $(\varepsilon_x, \varepsilon_y, \gamma_{yz}, \gamma_{xz}, \gamma_{xy})$ and $(\sigma_x, \sigma_y, \tau_{yz}, \tau_{xz}, \tau_{xy})$ represent the strain and stress components, respectively. Stiffness coefficients $C_{ij}$ can be written by employing the material properties in Eq. (6) as follows:

$$C_{11} = C_{22} = \frac{E(z,T)}{1-\nu^2(z,T)}, C_{12} = \nu(z,T)C_{11}, C_{44} = C_{55} = C_{66} = \frac{E(z,T)}{2(1+\nu(z,T))}, \tag{9}$$

## 2.3 Equations of motion

The strain energy of the functionally graded plate (FGP) is expressed as follows:

$$U = U_p + U_T \tag{10}$$

| | Material | $P_{-1}$ | $P_0$ | $P_1$ | $P_2$ | $P_3$ |
|---|---|---|---|---|---|---|
| $E$ | $SI_3N_4$ | 0 | $348.43e^{+9}$ | $-3.070e^{-4}$ | $2.160e^{-7}$ | $-8.946e^{-11}$ |
| | SUS304 | | $201.04e^{+9}$ | $3.079e^{-3}$ | $-6.534e^{-7}$ | 0 |
| | $ZrO_2$ | | $244.27e^{+9}$ | $-1.371e^{-3}$ | $1.214e^{-6}$ | $-3.681e^{-10}$ |
| | Ti-6Al-4 V | | $122.56e^{+9}$ | $-4.586e^{-4}$ | 0 | 0 |
| $\nu$ | $SI_3N_4$ | | 0.2400 | 0 | | |
| | SUS304 | | 0.3262 | $-2.002e^{-4}$ | $3.797e^{-7}$ | |
| | $ZrO_2$ | | 0.3330 | 0 | 0 | |
| | Ti-6Al-4 V | | 0.2888 | $1.108e^{-4}$ | | |
| $\alpha$ | $SI_3N_4$ | | $5.8723e^{-6}$ | $9.095e^{-6}$ | | |
| | SUS304 | | $12.330e^{-6}$ | $8.086e^{-6}$ | | |
| | $ZrO_2$ | | $12.766e^{-6}$ | $-1.491e^{-3}$ | $1.006e^{-5}$ | $-6.778e^{-11}$ |
| | Ti-6Al-4 V | | $7.5788e^{-6}$ | $6.638e^{-4}$ | $-3.147e^{-6}$ | 0 |
| $\rho$ | $SI_3N_4$ | | 2370 | 0 | 0 | |
| | SUS304 | | 8166 | | | |
| | $ZrO_2$ | | 3000 | | | |
| | Ti-6Al-4 V | | 4429 | | | |
| $k$ | $SI_3N_4$ | | 9.19 | | | |
| | SUS304 | | 12.04 | | | |
| | $ZrO_2$ | | 1.80 | | | |
| | Ti-6Al-4 V | | 7.82 | | | |

**Table 1.**
*Temperature-dependent coefficients for $ZrO_2$/Ti-6Al-4 V and $SI_3N_4$/SUS304.*

where $U_T$ and $U_p$ represent the strain energies due to thermal and mechanical effects, respectively. These components can be formulated as follows, incorporating temperature-dependent material properties and strain–displacement relationships, as reported in [36–38]

$$U_p = \frac{1}{2}\int_V \left[\sigma_x\,\varepsilon_x + \sigma_y\,\varepsilon_y + \tau_{xy}\,\gamma_{xy} + \tau_{yz}\,\gamma_{yz} + \tau_{xz}\,\gamma_{xz}\right]dV \tag{11}$$

$$U_T = \frac{1}{2}\int_V \left[\sigma_x^T d_{11} + \sigma_y^T d_{22}\right]dV \tag{12}$$

where $d_{ij}$ $(i,j = 1,2)$ denotes the non-linear relationship of strain–displacement [38]. By putting $d_{ij}$ in Eqs. (11) and (12) can be determined:

$$U_T = \frac{1}{2}\int_V \left\{\sigma_x^T\left[\left(\frac{\partial u}{\partial x}\right)^2 + \left(\frac{\partial v}{\partial x}\right)^2 + \left(\frac{\partial w}{\partial x}\right)^2\right] + \sigma_y^T\left[\left(\frac{\partial u}{\partial y}\right)^2 + \left(\frac{\partial v}{\partial y}\right)^2 + \left(\frac{\partial w}{\partial y}\right)^2\right]\right\}dV \tag{13}$$

with

$$\sigma_x^T = -(C_{11} + C_{12})\alpha(z, T)\Delta T(z) \text{ and } \sigma_y^T = -(C_{22} + C_{12})\alpha(z, T)\Delta T(z) \tag{14}$$

The kinetic energy of the functionally graded plate (FGP) is expressed as

$$K = \frac{1}{2}\int_V \rho(z, T)[\dot{u} + \dot{v} + \dot{w}]dV \tag{15}$$

where $\rho(z, T)$ denotes the mass density, which varies through the thickness of the plate according to the material gradation. The terms $\dot{u}, \dot{v},$ and $\dot{w}$ represent the time derivatives of the displacements in the $x, y,$ and $z$ directions, respectively. This formulation accounts for the dynamic response of the FGP by integrating over the entire volume V of the plate.

The change of external energy due to the foundation's reaction force can be given as follows:

$$\delta V = \int_A \left(k_w - k_p \nabla^2 + c_t \frac{\partial}{\partial t}\right) w(x, y)\delta w dA \tag{16}$$

The parameter $k_w$ denotes the Winkler foundation stiffness, while $c_t$ and $k_p$ represent the viscosity parameter and the shear layer foundation stiffness, respectively. These parameters characterize the viscoelastic foundation supporting the functionally graded plate (FGP), influencing its dynamic and static responses under external loads.

For an elastic body, Hamilton's principle, which governs the motion of mechanical systems, is expressed as [39–41]

$$\int_{t_1}^{t_2} (\delta U - \delta K + \delta V)dt = 0 \tag{17}$$

Putting Eq. (2) into Eq. (8) and using Eqs. (1) and (17), the motion equations for FG plate are expressed as follows:

$$\begin{aligned}(A_{11} + A_{11}^T)d_{11}u_0 + (A_{66} + A_{22}^T)d_{22}u_0 + (A_{12} + A_{66})d_{12}v_0 - (B_{11} + B_{11}^T)d_{111}w_b \\ -(B_{12} + 2B_{66} + B_{22}^T)d_{122}w_b = I_0\ddot{u}_0 - I_1 d_1\ddot{w}_b,\end{aligned} \tag{18}$$

$$\begin{aligned}(A_{22} + A_{22}^T)d_{22}v_0 + (A_{66} + A_{11}^T)d_{11}v_0 + (A_{12} + A_{66})d_{12}u_0 - (B_{22} + B_{22}^T)d_{222}w_b \\ -(B_{12} + 2B_{66} + B_{11}^T)d_{112}w_b = I_0\ddot{v}_0 - I_1 d_2\ddot{w}_b,\end{aligned} \tag{19}$$

$$\begin{aligned}&(B_{11} + B_{11}^T)d_{111}u_0 + (B_{12} + 2B_{66} + B_{22}^T)d_{122}u_0 + (B_{12} + 2B_{66} + B_{11}^T)d_{112}v_0 + (B_{22} + B_{22}^T)d_{222}v_0 \\ &-(D_{11} + D_{11}^T)d_{1111}w_b - 2(2D_{66} + D_{12})d_{1122}w_b - (D_{22} + D_{22}^T)d_{2222}w_b + A_{11}^T(d_{11}w_s + d_{11}w_b) \\ &+A_{22}^T(d_{22}w_b + d_{22}w_s) - (D_{11}^T + D_{22}^T)d_{1122}w_b - k_w(w_b + w_s) + k_p\nabla^2(w_b + w_s) \\ &-c_t\frac{\partial}{\partial t}(w_b + w_s) = I_0(\ddot{w}_b + \ddot{w}_s) + I_1(d_1\ddot{u}_0 + d_2\ddot{v}_0) - I_2(d_{11}\ddot{w}_b + d_{22}\ddot{w}_b)\end{aligned} \tag{20}$$

$$\begin{gathered}A_{44}^{s}d_{22}w_{s}+A_{55}^{s}d_{11}w_{s}+A_{11}^{T}(d_{11}w_{s}+d_{11}w_{b})+A_{22}^{T}(d_{22}w_{b}+d_{22}w_{s})\\ -k_{w}(w_{b}+w_{s})+k_{p}\nabla^{2}(w_{b}+w_{s})-c_{t}\frac{\partial}{\partial t}(w_{b}+w_{s})=I_{0}(\ddot{w}_{b}+\ddot{w}_{s})+J_{1}(d_{1}\ddot{u}_{0}+d_{2}\ddot{v}_{0})\\ -J_{2}(d_{11}\ddot{w}_{b}+d_{22}\ddot{w}_{b})\end{gathered} \tag{21}$$

where $d_{ijlm}$, $d_{ijl}$, and $d_{ij}$ denote the differential operators:

$$d_{ij}=\frac{\partial^{2}}{\partial x_{i}\partial x_{j}},d_{ijl}=\frac{\partial^{3}}{\partial x_{i}\partial x_{j}\partial x_{l}},d_{ijlm}=\frac{\partial^{4}}{\partial x_{i}\partial x_{j}\partial x_{l}\partial x_{m}},d_{i}=\frac{\partial}{\partial x_{i}},(i,j,l,m=1,2). \tag{22}$$

and stiffness components can be expressed as follows:

$$\begin{Bmatrix}A_{11} & B_{11} & D_{11}\\ A_{12} & B_{12} & D_{12}\\ A_{66} & B_{66} & D_{66}\end{Bmatrix}=\int_{-h/2}^{h/2}C_{11}(1,z,z^{2})\begin{Bmatrix}1\\ \nu\\ \frac{1-\nu}{2}\end{Bmatrix}dz, \tag{23}$$

$$(A_{22},B_{22},D_{22})=(A_{11},B_{11},D_{11}), \tag{24}$$

$$A_{44}^{s}=A_{55}^{s}=\int_{-h/2}^{h/2}C_{44}[g(z)]dz, \tag{25}$$

$$\begin{Bmatrix}A_{11}^{T} & B_{11}^{T} & D_{11}^{T}\\ A_{22}^{T} & B_{22}^{T} & D_{22}^{T}\end{Bmatrix}=\int_{-h/2}^{h/2}(1,z,z^{2})\begin{Bmatrix}\sigma_{x}^{T}\\ \sigma_{y}^{T}\end{Bmatrix}dz \tag{26}$$

The inertias can be given as follows:

$$(I_{1},I_{2},I_{3})=\int_{-h/2}^{h/2}(1,z,z^{2})\rho(z)dz \tag{27}$$

## 2.4 Temperature field

### *2.4.1 Uniform temperature*

A uniform temperature field across the thickness of the functionally graded plate (FGP) can be described using the following equation:

$$T(z)=T_{0}+\Delta T(z) \tag{28}$$

where $T(z)$ represents the temperature at a given position zzz along the thickness direction, $T_0 = 300\ K$ is the room temperature (or reference temperature), and $\Delta T(z)$ is the constant temperature increase applied uniformly throughout the plate.

This assumption simplifies thermal analysis by considering that temperature effects are evenly distributed, which is particularly useful for analyzing thermal stresses and deformations in homogeneous or mildly graded plates.

### 2.4.2 Linear temperature

When the temperatures at the bottom and top surfaces of the plate are denoted by $T_b$ and $T_t$, respectively, and a linear temperature gradient is assumed across the thickness, the temperature field is expressed as follows:

$$T(z) = T_0 + \Delta T\left(\frac{z}{h} + \frac{1}{2}\right) \tag{29}$$

The temperature gradient is given by $\Delta T = T_t - T_b$. The room temperature $T_0$ can be taken to be 300 $K$.

This formulation captures a linearly varying temperature profile, commonly used to model thermal gradients in functionally graded materials. It allows for an accurate analysis of thermal stresses and deformations caused by non-uniform heating or cooling between the plate's surfaces.

### 2.4.3 Non-linear temperature

The steady-state heat conduction equation along the $z$-direction for the functionally graded plate (FGP) is expressed as

$$-\frac{d}{dz}\left(k(z)\frac{dT}{dz}\right) = 0 \tag{30}$$

With the boundary condition of $T(-h/2) = T_b = T_0$ and $T(h/2) = T_t$, a stress-free state is supposed to appear at room temperature ($T_0 = 300\ K$). The analytical solution to Eq. (29) is

$$T(z) = T_b - (T_t - T_b)\frac{\int_{-h/2}^{z} \frac{1}{k(z)}dz}{\int_{-h/2}^{h/2} \frac{1}{k(z)}dz}. \tag{31}$$

With the application of a power-law distribution for the functionally graded plate (FGP), the solution to Eq. (29) can be expressed using a polynomial series, as detailed in [36]:

$$T(z) = T_b + \frac{(T_t - T_b)}{C_{tb}}\left[\left(\frac{2z+h}{2h}\right) - \frac{k_{tb}}{(p+1)k_b}\left(\frac{2z+h}{2h}\right)^{p+1} + \frac{k_{tb}^2}{(2p+1)k_b^2}\left(\frac{2z+h}{2h}\right)^{2p+1} - \frac{k_{tb}^3}{(3p+1)k_b^3}\left(\frac{2z+h}{2h}\right)^{3p+1} + \frac{k_{tb}^4}{(4p+1)k_b^4}\left(\frac{2z+h}{2h}\right)^{4p+1} - \frac{k_{tb}^5}{(5p+1)k_b^5}\left(\frac{2z+h}{2h}\right)^{5p+1}\right] \tag{32}$$

in which

$$C_{tb} = 1 - \frac{k_{tb}}{(p+1)k_b} + \frac{k_{tb}^2}{(2p+1)k_b^2} - \frac{k_{tb}^3}{(3p+1)k_b^3} + \frac{k_{tb}^4}{(4p+1)k_b^4} - \frac{k_{tb}^5}{(5p+1)k_b^5} \tag{33}$$

with $k_{tb} = k_t - k_b$. $k_b$ and $k_t$ represent the thermal conductivity of the plate's bottom and top faces.

*2.4.4 Sinusoidal temperature rise*

Under sinusoidal temperature increase, the temperature field along the plate's thickness is considered as

$$T(z) = (T_t - T_b)\left[1 - \cos\left(\frac{\pi z}{2h} + \frac{\pi}{4}\right)\right] + T_b \tag{34}$$

## 3. Analytical solutions

Using the Navier approach [42], the displacement fields for a simply supported functionally graded plate (FGP) can be expressed as

$$\begin{Bmatrix} u_0 \\ v_0 \\ w_b \\ w_s \end{Bmatrix} = \sum_{m=1}^{\infty}\sum_{n=1}^{\infty} \begin{Bmatrix} U_{mn} e^{i\omega t} \cos(\lambda x)\sin(\mu y) \\ V_{mn} e^{i\omega t} \sin(\lambda x)\cos(\mu y) \\ W_{bmn} e^{i\omega t} \sin(\lambda x)\sin(\mu y) \\ W_{smn} e^{i\omega t} \sin(\lambda x)\sin(\mu y) \end{Bmatrix} \tag{35}$$

where $U_{mn}$, $V_{mn}$, $W_{bmn}$ and $W_{smn}$ are parameters to be calculated. $\omega$ denotes the eigenfrequency related to $(m, n)$ of the eigenmode, and $\lambda = m\pi/a$ and $\mu = n\pi/b$.

Putting Eq. (33) into motion Eqs. (18)–(21), the frequency equation can be expressed as follows:

$$\begin{bmatrix} a_{11} & a_{12} & a_{13} & a_{14} \\ a_{12} & a_{22} & a_{23} & a_{24} \\ a_{13} & a_{23} & a_{33} & a_{34} \\ a_{14} & a_{24} & a_{34} & a_{44} \end{bmatrix} - \omega^2 \begin{bmatrix} m_{11} & 0 & m_{13} & m_{14} \\ 0 & m_{22} & m_{23} & m_{24} \\ m_{13} & m_{23} & m_{33} & m_{34} \\ m_{14} & m_{24} & m_{34} & m_{44} \end{bmatrix} = \begin{Bmatrix} 0 \\ 0 \\ 0 \\ 0 \end{Bmatrix} \tag{36}$$

## 4. Numerical analysis and validation

In this section, various numerical examples are presented to validate the accuracy and effectiveness of the proposed model for analyzing the free vibration of temperature-dependent functionally graded plates (FGPs) resting on a viscoelastic foundation. The material gradation assumes that the bottom surface is metal-rich while the top surface is ceramic-rich, consistent with previous studies [27, 36, 43].

Validation is performed by comparing the results of the proposed model with established data from the literature. Different parameters, such as the power-law index, temperature field, and viscoelastic foundation properties, are considered to evaluate the dynamic behavior of the FGPs. The analysis demonstrates excellent agreement with benchmark results, confirming the reliability of the developed model under various thermal and mechanical conditions.

Confirmation is conceded by considering the values of different quantities as: $a = b = 0.2$ m, $h = 0.025$ m, $\nu = 0.3$. The same value of Poisson's ratio $\nu$ is supposed to be equally for metal and ceramic. However, as reported by [27, 36, 43], temperature-dependent relation is considered for Young's modulus and coefficient of thermal expansion of these materials.

The nondimensional frequency parameter is

$$\overline{\varpi} = \omega(a^2/h)\left[\rho_b(1-\nu^2)/E_b\right]^{1/2} \overline{k}_w = \frac{k_w a^4}{D_0}\ \overline{k}_p = \frac{k_p a^2}{D_0} \tag{37}$$

$$C_T = \frac{c_t a^4}{10^3 D_0}\ D_0 = E_b h^3/12(1-\nu^2) \tag{38}$$

where $E_b$ and $\rho_b$ are properties of materials at $T_0 = 300\ (K)$.

**Table 2** presents the dimensionless fundamental frequencies of $ZrO_2$/Ti–6Al–4 V FGP without a visco-elastic foundation. The novel FSDT model is demonstrated by validating the predicted frequencies with the results obtained from the other theories [27, 36, 43]. In **Table 2**, Shahrjerdi et al. [36] utilized a second-order shear deformation theory (SSDT) with seven variables; Huang and Shenn [43] worked with an HSDT with fewer variables than the SSDT, and a refined HSDT (four variables) is

| Mode (1,1) Natural frequency of FGP ($ZrO_2$ and Ti-6Al-4 V) | | $T_b = 300\ (K)$ | | | | |
|---|---|---|---|---|---|---|
| | | $T_t = 300\ (K)$ | $T_t = 400\ (K)$ | | $T_t = 600\ (K)$ | |
| | | | Temperature | | | |
| | | | Dependent | Independent | Dependent | Independent |
| $ZrO_2$ | Shahrjerdi et al. [36] | 8.333 | 7.614 | 7.892 | 5.469 | 6.924 |
| | Huang and Shen [43] | 8.273 | 7.868 | 8.122 | 6.685 | 7.686 |
| | Attia et al. [27] | 8.278 | 7.807 | 8.130 | 6.533 | 7.826 |
| | present | 8.263 | 7.792 | 8.114 | 6.517 | 7.810 |
| $p = 0.5$ | Shahrjerdi et al. [36] | 7.156 | 6.651 | 6.844 | 5.255 | 6.175 |
| | Huang and Shen [43] | 7.139 | 6.876 | 7.154 | 6.123 | 6.776 |
| | Attia et al. [27] | 7.111 | 6.781 | 7.005 | 5.931 | 6.789 |
| | present | 7.099 | 6.769 | 6.993 | 5.918 | 6.776 |
| $p = 1$ | Shahrjerdi et al. [36] | 6 .700 | 6.281 | 6.446 | 5.167 | 5.904 |
| | Huang and Shen [43] | 6.657 | 6.437 | 6.592 | 5.819 | 6.362 |
| | Attia et al. [27] | 6.657 | 6.375 | 6.565 | 5.664 | 6.378 |
| | present | 6.645 | 6.363 | 6.553 | 5.653 | 6.366 |
| $p = 2$ | Shahrjerdi et al. [36] | 6.333 | 5.992 | 6.132 | 5.139 | 5.711 |
| | Huang and Shen [43] | 6.286 | 6.101 | 6.238 | 5.612 | 6.056 |
| | Attia et al. [27] | 6.287 | 6.047 | 6.208 | 5.467 | 6.049 |
| | present | 6.275 | 6.035 | 6.196 | 5.455 | 6.036 |
| Ti-6Al-4 V | Shahrjerdi et al. [36] | 5.439 | 5.103 | 5.333 | 4.836 | 5.115 |
| | Huang and Shen [43] | 5.400 | 5.322 | 5.389 | 5.118 | 5.284 |
| | Attia et al. [27] | 5.403 | 5.303 | 5.361 | 5.132 | 5.275 |
| | present | 5.393 | 5.293 | 5.351 | 5.122 | 5.265 |

**Table 2.**
*Comparison of nondimensional natural frequency parameter for ($ZrO_2$/Ti-6Al-4 V) FG plate with simply supported boundary condition under thermal environments and without elastic foundation.*

used by Attia et al. [27]. The validation indicates that the present theory is working correctly and agrees with the results obtained from previously published works cited above, particularly those obtained by Huang and Shen [43] and Attia et al. [27] for every value of the power law index $p$, regardless of temperature-independent or temperature dependent of FGPs.

This section focuses on the dynamic analysis of functionally graded plates (FGPs), specifically $Si_3N_4$/SUS304 plates, to evaluate the accuracy and applicability of the developed theoretical model. The material gradation assumes a ceramic-rich top surface and a metal-rich bottom surface, with the Poisson's ratio set at 0.28.

### 4.1 Fundamental frequency analysis

The dimensionless fundamental frequencies for different values of the power-law index $p$, which governs the material distribution across the plate thickness, are computed and presented in **Table 3**. These results are compared with those obtained from

| Mode (1,1) Natural frequency of FGP ($Si_3N_4$ and SUS304) | | $T_b = 300\ (K)$ | | | | |
|---|---|---|---|---|---|---|
| | | $T_t = 300\ (K)$ | $T_t = 400\ (K)$ | | $T_t = 600\ (K)$ | |
| | | | Temperature | | | |
| | | | Dependent | Independent | Dependent | Independent |
| $Si_3N_4$ | Shahrjerdi et al. [36] | 12.506 | 12.175 | 12.248 | 11.461 | 11.716 |
| | Huang and Shen [43] | 12.495 | 12.397 | 12 .382 | 11.984 | 12.213 |
| | Attia et al. [27] | 12.507 | 12.307 | 12.376 | 11.886 | 12.113 |
| | present | 12.484 | 12.284 | 12.354 | 11.863 | 12.090 |
| $p = 0.5$ | Shahrjerdi et al. [36] | 8.652 | 8.361 | 8.405 | 7.708 | 7.887 |
| | Huang and Shen [43] | 8.675 | 8.615 | 8.641 | 8.269 | 8.425 |
| | Attia et al. [27] | 8.609 | 8.453 | 8.498 | 8.117 | 8.272 |
| | present | 8.595 | 8.439 | 8.484 | 8.103 | 8.257 |
| $p = 1$ | Shahrjerdi et al. [36] | 7.584 | 7.306 | 7.342 | 6.674 | 6.834 |
| | Huang and Shen [43] | 7.555 | 7.474 | 7.514 | 7.171 | 7.305 |
| | Attia et al. [27] | 7.544 | 7.399 | 7.437 | 7.082 | 7.217 |
| | present | 7.531 | 7.385 | 7.423 | 7.068 | 7.203 |
| $p = 2$ | Shahrjerdi et al. [36] | 6.811 | 6.545 | 6.575 | 5.929 | 6.077 |
| | Huang and Shen [43] | 6.777 | 6.693 | 6.728 | 6.398 | 6.523 |
| | Attia et al. [27] | 6.771 | 6.631 | 6.664 | 6.323 | 6.447 |
| | present | 6.758 | 6.618 | 6.651 | 6.309 | 6.433 |
| SUS304 | Shahrjerdi et al. [36] | 5.410 | 5.161 | 5.178 | 4.526 | 4.682 |
| | Huang and Shen [43] | 5.405 | 5.311 | 5.335 | 4.971 | 5.104 |
| | Attia et al. [27] | 5.410 | 5.272 | 5.295 | 4.922 | 5.055 |
| | present | 5.400 | 5.268 | 5.289 | 4.934 | 5.061 |

**Table 3.**
*Comparison of nondimensional natural frequency parameter for ($Si_3N_4$/SUS304) FG plate with simply supported boundary condition under thermal environments.*

other established theories, such as those of Huang and Shen [43] and Attia et al. [27]. The comparison reveals remarkable agreement, demonstrating the validity and precision of the present theory in capturing the dynamic behavior of FGPs. The alignment of results indicates that the developed approach effectively incorporates the influence of material gradation and thermal effects on the natural frequencies.

## 4.2 Analysis of other material combinations

To further validate the model, numerical results for $ZrO_2$/Ti–6Al–4 V and $Si_3N_4$/SUS304 plates are presented in **Tables 4** and **5**, respectively. These results are benchmarked against data from Huang and Shen [43], Shahrjerdi et al. [36], and Attia et al. [27]. Across different mode shapes and material configurations, the proposed model consistently produces results that closely match the reference data. This consistency highlights the robustness and versatility of the theory for analyzing a wide range of material combinations and boundary conditions.

| Mode numbers of FGP ($ZrO_2$ and Ti-6Al-4 V) | | $T_b = 300\ (K)$ | | | | |
|---|---|---|---|---|---|---|
| | | $T_t = 300\ (K)$ | $T_t = 400\ (K)$ | | $T_t = 600\ (K)$ | |
| | | | Temperature | | | |
| | | | Dependent | Independent | Dependent | Independent |
| (1,1) | Shahrjerdi et al. [36] | 6.333 | 5.992 | 6.132 | 5.139 | 5.711 |
| | Huang and Shen [43] | 6.286 | 6.101 | 6.238 | 5.612 | 6.056 |
| | Attia et al. [27] | 6.287 | 6.047 | 6.208 | 5.467 | 6.049 |
| | present | 6.275 | 6.035 | 6.196 | 5.455 | 6.036 |
| (1,2) | Shahrjerdi et al. [36] | 14.896 | 14.383 | 14.684 | 13.260 | 14.253 |
| | Huang and Shen [43] | 14.625 | 14.372 | 14.655 | 13.611 | 14.474 |
| | Attia et al. [27] | 14.665 | 14.265 | 14.581 | 13.416 | 14.412 |
| | present | 14.602 | 14.203 | 14.517 | 13.353 | 14.348 |
| (2,2) | Shahrjerdi et al. [36] | 22.608 | 21.942 | 22.386 | 20.557 | 21.935 |
| | Huang and Shen [43] | 21.978 | 21.653 | 22.078 | 20.652 | 21.896 |
| | Attia et al. [27] | 22.123 | 21.584 | 22.034 | 20.489 | 21.855 |
| | present | 21.984 | 21.449 | 21.895 | 20.359 | 21.715 |
| (1,3) | Shahrjerdi et al. [36] | 27.392 | 26.630 | 27.163 | 25.077 | 26.700 |
| | Huang and Shen [43] | 26.454 | 26.113 | 26.605 | 24.961 | 26.435 |
| | Attia et al. [27] | 26.704 | 26.081 | 26.612 | 24.837 | 26.427 |
| | present | 26.506 | 25.889 | 26.414 | 24.653 | 26.228 |
| (2,3) | Shahrjerdi et al. [36] | 34.106 | 33.211 | 33.867 | 31.425 | 33.384 |
| | Huang and Shen [43] | 32.659 | 32.239 | 32.840 | 30.904 | 32.664 |
| | Attia et al. [27] | 33.109 | 32.371 | 33.013 | 30.920 | 32.819 |
| | present | 32.813 | 32.084 | 32.716 | 30.644 | 32.521 |

**Table 4.**
*Comparison of nondimensional natural frequency parameter for (ZrO2/Ti-6Al-4 V) FG plate with simply supported boundary condition under thermal environments ($p = 2$).*

| Mode numbers of FGP ($Si_3N_4$/SUS304) | | $T_b = 300\ (K)$ | | | | |
|---|---|---|---|---|---|---|
| | | $T_t = 300\ (K)$ | $T_t = 400\ (K)$ | | $T_t = 600\ (K)$ | |
| | | | Temperature | | | |
| | | | Dependent | Independent | Dependent | Independent |
| (1,1) | [36] | 6.811 | 6.445 | 6.575 | 5.929 | 6.077 |
| | [43] | 6.777 | 6.693 | 6.728 | 6.398 | 6.523 |
| | [27] | 6.770 | 6.631 | 6.664 | 6.323 | 6.447 |
| | present | 6.758 | 6.618 | 6.651 | 6.309 | 6.433 |
| (1,2) | [36] | 16.017 | 15.708 | 15.769 | 15.002 | 15.262 |
| | [43] | 15.809 | 15.762 | 15.836 | 15.384 | 15.632 |
| | [27] | 15.812 | 15.628 | 15.699 | 15.229 | 15.472 |
| | present | 15.745 | 15.561 | 15.632 | 15.161 | 15.403 |
| (2,2) | [36] | 24.307 | 23.958 | 24.047 | 23.154 | 23.517 |
| | [43] | 23.806 | 23.786 | 23.893 | 23.327 | 23.685 |
| | [27] | 23.874 | 23.652 | 23.755 | 23.167 | 23.517 |
| | present | 23.727 | 23.504 | 23.607 | 23.019 | 23.367 |
| (1,3) | [36] | 29.446 | 29.071 | 29.177 | 28.204 | 28.632 |
| | [43] | 28.687 | 28.686 | 28.816 | 28.185 | 28.609 |
| | [27] | 28.831 | 28.586 | 28.709 | 28.049 | 28.463 |
| | present | 28.621 | 28.375 | 28.498 | 27.838 | 28.249 |
| (2,3) | [36] | 36.657 | 36.247 | 36.376 | 35.290 | 35.809 |
| | [43] | 35.466 | 35.491 | 35.648 | 34.918 | 35.436 |
| | [27] | 35.768 | 35.489 | 35.640 | 34.879 | 35.383 |
| | present | 35.452 | 35.173 | 35.323 | 34.563 | 35.063 |

**Table 5.**
*Comparison of nondimensional natural frequency parameter for ($Si_3N_4$/SUS304) FG plate with simply supported boundary conditions under thermal environments ($p = 2$).*

### 4.3 Influence of key parameters

The dynamic response of FGPs is influenced by several parameters, including thermal fields, power-law index $p$, and the properties of the elastic foundation. The following observations are made:

- *Power-law index*: As the power-law index $p$ increases, indicating a greater concentration of ceramic material near the top surface, the natural frequencies increase. This is attributed to the higher stiffness of the ceramic phase, which enhances the plate's rigidity.

- *Thermal effects*: Increasing the temperature difference between the bottom and top surfaces results in a noticeable decrease in the natural frequencies. This behavior reflects the softening effect induced by thermal gradients, which reduce the overall stiffness of the plate.

- *Temperature dependence*: When considering temperature-dependent material properties, the natural frequencies are observed to be lower than those for temperature-independent FGPs. This demonstrates the critical importance of accounting for temperature dependency in material properties to accurately predict the dynamic behavior under thermal loads.

- *Elastic foundation*: The viscoelastic foundation significantly impacts the dynamic response, with the Winkler parameter, shear layer stiffness, and viscosity playing crucial roles in modifying the frequencies. The foundation provides additional stiffness and damping, which can either enhance or mitigate the frequency reduction caused by thermal effects, depending on its properties.

### 4.4 Practical implications

The insights gained from these analyses are essential for the design and optimization of FGPs in engineering applications, particularly in environments with complex thermal and mechanical loading conditions. For example, the findings are directly relevant to the aerospace, automotive, and energy sectors, where components are often exposed to temperature gradients and require high-performance materials.

By capturing the interplay between material gradation, thermal effects, and foundation properties, the proposed model provides a comprehensive framework for evaluating the vibrational behavior of FGPs. This enables engineers and researchers to make informed decisions in designing reliable and efficient structures.

### 4.5 Effect of elastic foundation on natural frequencies

#### *4.5.1 Influence of foundation coefficients*

**Table 6** highlights the impact of the elastic foundation coefficients, specifically the Winkler ($kw$) and Pasternak ($kp$) parameters, on the natural frequencies of functionally graded plates (FGPs). For this analysis, the damping coefficient ($c_{t,}$) is set to zero. It is observed that the natural frequency increases with the rise in either $kw$ or $kp$. This behavior can be attributed to the additional stiffness provided by the elastic foundation, which effectively enhances the structural rigidity of the plate.

The results indicate that the highest natural frequency values are obtained when the plate rests on a combined Winkler-Pasternak foundation. This combination provides both normal and shear stiffness contributions, offering superior support and resistance to deformation compared to a foundation with only Winkler or Pasternak components.

#### *4.5.2 Detailed analysis of foundation parameters and damping*

**Tables 7** and **8** further explore the influence of the Winkler parameter ($kw$), the Pasternak parameter ($kp$), and the damping coefficient ($c_{t,}$) on the natural frequencies. The following observations are made:

- *Winkler parameter (kw)*: An increase in kwk_wkw results in a significant rise in the natural frequency. This is because $kw$ models the normal stiffness provided by the foundation, directly counteracting the deformation of the plate and increasing its dynamic response.

| ($ZrO_2$ and Ti-6Al-4 V) | | $T_b = 300\ (K)$ | | | | |
|---|---|---|---|---|---|---|
| | | $T_t = 300\ (K)$ | $T_t = 400\ (K)$ | | $T_t = 600\ (K)$ | |
| | | | Temperature | | | |
| | | | Dependent | Independent | Dependent | Independent |
| $\bar{k}_w = 100$ $\bar{k}_p = 0$ | $ZrO_2$ | 8.962 | 8.529 | 8.825 | 7.383 | 8.546 |
| | $p = 0.5$ | 7.796 | 7.497 | 7.699 | 6.739 | 7.503 |
| | $p = 1$ | 7.338 | 7.086 | 7.256 | 6.455 | 7.088 |
| | $p = 2$ | 6.964 | 6.750 | 6.894 | 6.237 | 6.750 |
| | Ti-6Al-4 V | 6.103 | 5.997 | 6.056 | 5.828 | 5.961 |
| $\bar{k}_w = 100$ $\bar{k}_p = 100$ | $ZrO_2$ | 17.832 | 17.619 | 17.763 | 17.092 | 17.627 |
| | $p = 0.5$ | 16.298 | 16.159 | 16.252 | 15.823 | 16.160 |
| | $p = 1$ | 15.670 | 15.555 | 15.632 | 15.281 | 15.554 |
| | $p = 2$ | 15.125 | 15.029 | 15.092 | 14.808 | 15.027 |
| | Ti-6Al-4 V | 14.076 | 14.033 | 14.059 | 13.961 | 14.018 |

**Table 6.**
*Comparison of nondimensional natural frequency parameter for ($ZrO_2$/Ti-6Al-4 V) FG plate with simply supported boundary conditions under thermal environments and resting on elastic foundations.*

| Mode (1,1) Natural frequency of FGP ($ZrO_2$ and Ti-6Al-4 V) | | $T_b = 300\ (K)$ | | | | |
|---|---|---|---|---|---|---|
| | | $T_t = 300\ (K)$ | $T_t = 400\ (K)$ | | $T_t = 600\ (K)$ | |
| | | | Temperature | | | |
| | | | Dependent | Independent | Dependent | Independent |
| $\bar{C}_T = 0$ | $ZrO_2$ | 8.962 | 8.529 | 8.825 | 7.383 | 8.546 |
| | $p = 0.5$ | 7.796 | 7.497 | 7.699 | 6.739 | 7.503 |
| | $p = 1$ | 7.338 | 7.086 | 7.256 | 6.455 | 7.088 |
| | $p = 2$ | 6.964 | 6.750 | 6.894 | 6.237 | 6.750 |
| | Ti-6Al-4 V | 6.103 | 5.997 | 6.056 | 5.828 | 5.961 |
| $\bar{C}_T = 0.1$ | $ZrO_2$ | 9.292 | 8.876 | 8.994 | 7.779 | 8.721 |
| | $p = 0.5$ | 8.122 | 7.791 | 7.849 | 7.114 | 7.619 |
| | $p = 1$ | 7.663 | 7.836 | 7.404 | 6.822 | 7.202 |
| | $p = 2$ | 7.285 | 7.081 | 7.043 | 6.593 | 6.872 |
| | Ti-6Al-4 V | 6.428 | 6.328 | 6.384 | 6.168 | 6.294 |
| $\bar{C}_T = 0.2$ | $ZrO_2$ | 9.610 | 9.209 | 9.483 | 8.158 | 9.224 |
| | $p = 0.5$ | 8.435 | 8.160 | 8.329 | 7.469 | 8.115 |
| | $p = 1$ | 7.974 | 7.741 | 7.881 | 7.169 | 7.691 |
| | $p = 2$ | 7.592 | 7.396 | 7.513 | 6.931 | 7.354 |
| | Ti-6Al-4 V | 6.738 | 6.643 | 6.695 | 6.490 | 6.609 |

**Table 7.**
*Comparison of nondimensional natural frequency parameter for ($ZrO_2$/Ti-6Al-4 V) FG plate with simply supported boundary conditions under thermal environments and resting on a viscoelastic foundation* $\bar{k}_w = 100, \bar{k}_p = 0$.

| Mode (1,1) Natural frequency of FGP ($ZrO_2$ and Ti-6Al-4 V) | | $T_b = 300\ (K)$ | | | | |
|---|---|---|---|---|---|---|
| | | $T_t = 300\ (K)$ | $T_t = 400\ (K)$ | | $T_t = 600\ (K)$ | |
| | | | Temperature | | | |
| | | | Dependent | Independent | Dependent | Independent |
| $\overline{C}_T = 0$ | $ZrO_2$ | 17.832 | 17.619 | 17.763 | 17.092 | 17.627 |
| | $p = 0.5$ | 16.298 | 16.159 | 16.252 | 15.823 | 16.160 |
| | $p = 1$ | 15.670 | 15.555 | 15.632 | 15.281 | 15.554 |
| | $p = 2$ | 15.125 | 15.029 | 15.092 | 14.808 | 15.027 |
| | Ti-6Al-4 V | 14.076 | 14.033 | 14.059 | 13.961 | 14.018 |
| $\overline{C}_T = 0.2$ | $ZrO_2$ | 18.167 | 17.957 | 18.099 | 17.441 | 17.965 |
| | $p = 0.5$ | 16.613 | 16.477 | 16.568 | 16.148 | 16.478 |
| | $p = 1$ | 15.977 | 15.864 | 15.939 | 15.596 | 15.863 |
| | $p = 2$ | 15.423 | 15.329 | 15.392 | 15.113 | 15.328 |
| | Ti-6Al-4 V | 14.365 | 14.321 | 14.346 | 14.250 | 14.306 |

**Table 8.**
*Comparison of nondimensional natural frequency parameter for ($ZrO_2$/Ti-6Al-4 V) FG plate with simply supported boundary conditions under thermal environments and resting on a viscoelastic foundation* $\overline{k}_w = 100, \overline{k}_p = 100$.

- *Pasternak parameter (kp)*: The Pasternak parameter provides an additional layer of shear stiffness, further stabilizing the plate. The results show that increasing *kp* enhances the natural frequencies, especially when combined with a high *kp*. This indicates that the Pasternak foundation contributes to the plate's resistance against lateral shear deformations.

- *Damping coefficient ($c_{t,}$)*: When the damping coefficient $c_{t,}$ is included, its effect is more pronounced in the case of the Winkler-Pasternak foundation. The presence of damping introduces energy dissipation mechanisms that reduce the vibrational amplitude but slightly lower the natural frequencies depending on the damping magnitude. This behavior is essential for applications where vibration control and energy dissipation are critical.

### *4.5.3 Practical implications*

The analysis underscores the critical role of foundation parameters in the dynamic behavior of FGPs. A proper combination of Winkler and Pasternak stiffness coefficients can significantly enhance the structural performance of FGPs, making them more resilient to external disturbances. Additionally, the inclusion of damping provides valuable insights for applications requiring vibration isolation or energy dissipation.

Such findings are particularly relevant for designing advanced structures in civil engineering (e.g., foundation-supported slabs), aerospace (e.g., panels subjected to thermal and mechanical loads), and mechanical systems requiring precise dynamic performance under complex loading conditions.

The variation of the first four frequencies as a function of linear, non-linear, uniform, and sinusoidal temperature fields in simply supported FGP is plotted in **Figures 2–5**. The combination of $ZrO_2$/Ti- 6Al-4 V is assumed with geometric and material parameters of $p = 1$, $a = b = 0.2$ and $a/h = 10$. The nondimensional natural frequency parameter is considered to be $\overline{\omega} = \omega(b^2/\pi^2)[I_0/D_0]^{1/2}$, where $I_0 = \rho h$ and it is noted that $\rho$ and $\nu$ are selected to be the values of Ti-6Al-4 V calculated at the room temperature.

As expected, the natural frequencies of the functionally graded plate (FGP) decrease with increasing temperature due to the reduction of Young's modulus as the

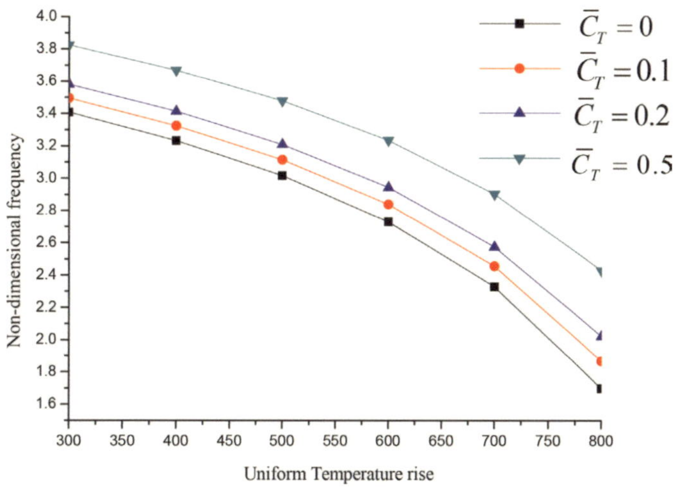


**Figure 2.**
*Influence of the damping coefficient on the nondimensional frequency parameters for uniform temperature field for ($ZrO_2$/Ti-6Al-4 V) FG plate with simply supported boundary condition when $a = 0.2$, $a/h = 10$ and $p = 1$.*

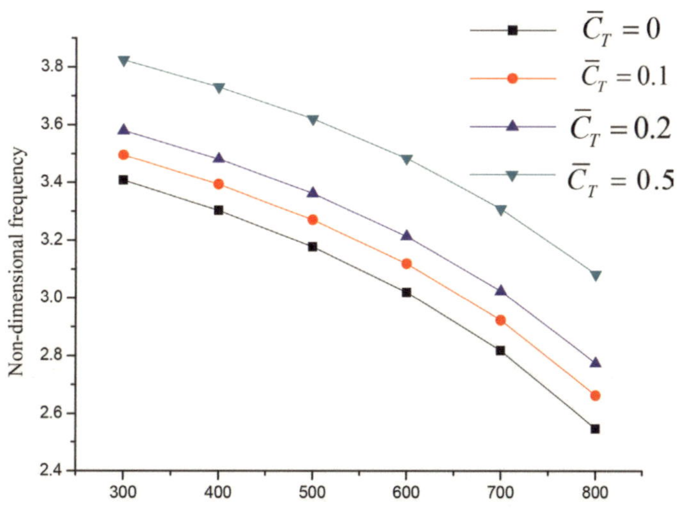


**Figure 3.**
*Influence of the damping coefficient on the nondimensional frequency parameters for linear temperature field for ($ZrO_2$/Ti-6Al-4 V) FG plate with simply supported boundary condition when $a = 0.2$, $a/h = 10$, and $p = 1$.*

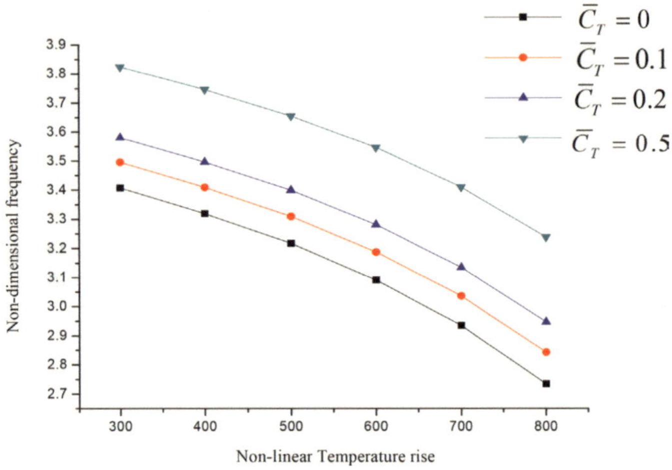


**Figure 4.**
*Influence of the damping coefficient on the nondimensional frequency parameters for non-linear temperature field for ($ZrO_2$/Ti-6Al-4 V) FG plate with simply supported boundary condition when $a = 0.2$, $a/h = 10$, and $p = 1$.*

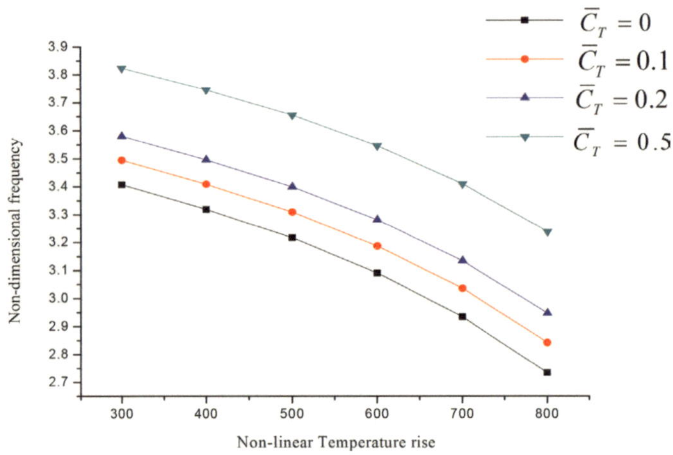


**Figure 5.**
*Influence of the damping coefficient on the nondimensional frequency parameters for sinusoidal temperature field for ($ZrO_2$/Ti-6Al-4 V) FG plate with simply supported boundary condition when $a = 0.2$, $a/h = 10$, and $p = 1$.*

material heats up. This behavior highlights the sensitivity of FGPs to thermal environments, where elevated temperatures reduce the stiffness of the plate, directly influencing its dynamic response.

The results, illustrated in **Figures 2–5**, show that the impact of temperature is more significant under uniform thermal conditions compared to linear, non-linear, or sinusoidal temperature fields. In uniform temperature conditions, the constant temperature increase across the plate leads to a uniform degradation of material properties, resulting in the most pronounced reduction in natural frequencies. Conversely, in non-linear and sinusoidal thermal profiles, the temperature variation across the thickness introduces localized stiffness gradients, which mitigate the overall reduction in frequency.

Additionally, the damping coefficient ($C_T$) plays a crucial role in the vibrational behavior of the plate. An increase in $C_T$ results in higher natural frequencies, as shown in **Figures 4** and **5**. This behavior can be attributed to the enhanced energy dissipation provided by the damping mechanism, which effectively counteracts the destabilizing effects of thermal loads. The increased damping makes the plate appear more rigid, improving its resistance to vibrations caused by external disturbances.

#### *4.5.4 Practical implications*

These findings underscore the importance of considering thermal conditions and damping effects when designing FGPs for high-temperature environments. For applications in aerospace, energy, and mechanical systems, where components are often subjected to varying thermal loads, optimizing the material composition and damping properties is critical to ensure structural stability and performance.

## 5. Conclusions

In this chapter, an efficient first-order shear deformation theory (FSDT) was developed and applied to analyze the free vibration of functionally graded plates (FGPs) subjected to various thermal environments, including linear, non-linear, uniform, and sinusoidal temperature fields. The plates were assumed to rest on a three-parameter viscoelastic foundation, allowing for a comprehensive investigation of the influence of elastic and damping properties on the dynamic behavior of FGPs.

One of the key strengths of the proposed theory lies in its simplicity and computational efficiency. By reducing the number of variables and equations of motion compared to conventional FSDT, the developed model provides a practical yet accurate framework for analyzing FGPs. This efficiency is particularly advantageous for large-scale problems or iterative design processes in engineering applications.

The study highlighted the importance of considering temperature-dependent material properties that vary across the plate thickness according to a power-law distribution. This realistic representation captured the complex interplay between thermal effects and material heterogeneity, enabling accurate predictions of the natural frequencies of FGPs. The analysis demonstrated that uniform temperature fields lead to the most significant reductions in natural frequencies due to the uniform degradation of material stiffness, whereas non-linear and sinusoidal temperature distributions result in more localized effects.

The viscoelastic foundation, characterized by Winkler and Pasternak stiffness coefficients along with a damping coefficient, was shown to have a significant impact on the vibrational response. The study revealed that an increase in these parameters enhances the plate's stiffness and modifies its dynamic behavior. In particular, the inclusion of damping was found to mitigate vibrational amplitudes and increase the plate's rigidity, offering an effective mechanism for vibration control in practical applications.

### 5.1 Practical relevance and broader implications

The findings presented in this chapter have direct implications for the design and optimization of FGPs in engineering systems, particularly those exposed to varying thermal and mechanical loads. Industries such as aerospace, automotive, and energy systems, where lightweight, thermally resistant, and high-performance materials are required, can benefit from the insights gained here. The ability to model and predict the behavior of FGPs with such accuracy ensures more efficient and reliable structural designs.

## Author details

Abdelmoumen Anis Bousahla
Djillali Liabes University, Sidi Bel Abbés, Algeria

*Address all correspondence to: bousahla.anis@gmail.com

## References

[1] Bousahla AA et al. Free vibration of functionally graded plates with temperature-dependent properties on a viscoelastic foundation. Structural Engineering and Mechanics. 2023;**86**(1): 1-16. DOI: 10.12989/sem.2023.86.1.001

[2] Kar VR, Panda SK. Non-linear thermomechanical deformation behaviour of P-FGM shallow spherical shell panel. Chinese Journal of Aeronautics. 2016;**29**(1):173-183. DOI: 10.1016/j.cja.2015.12.007

[3] Shokravi M. Buckling of sandwich plates with FG-CNT-reinforced layers resting on orthotropic elastic medium using Reddy plate theory. Steel and Composite Structures, An International Journal. 2017;**23**(6):623-631. DOI: 10.12989/SCS.2017.23.6.623

[4] Timoshenko S, Woinowsky-Krieger S. Theory of Plates and Shells. Vol. 2. New York: McGraw-hill; 1959. pp. 240-246

[5] Leissa AW. The free vibration of rectangular plates. Journal of Sound and Vibration. 1973;**31**(3):257-293. DOI: 10.1016/S0022-460X(73)80371-2

[6] Javaheri R, Eslami MR. Thermal buckling of functionally graded plates. AIAA Journal. 2002;**40**(1):162-169. DOI: 10.2514/2.1626

[7] Mohammadi M, Saidi AR, Jomehzadeh E. Levy solution for buckling analysis of functionally graded rectangular plates. Applied Composite Materials. 2010;**17**(2):81-93. DOI: 10.1007/s10443-009-9100-z

[8] Hu Y, Zhang X. Parametric vibrations and stability of a functionally graded plate. Mechanics Based Design of Structures. 2011;**39**(3):367-377. DOI: 10.1080/15397734.2011.557970

[9] Ghannadpour SAM, Ovesy HR, Nassirnia M. Buckling analysis of functionally graded plates under thermal loadings using the finite strip method. Computers and Structures. 2012; **108-109**:93-99. DOI: 10.1016/j.compstruc.2012.02.011

[10] Chakraverty S, Pradhan KK. Free vibration of functionally graded thin rectangular plates resting on Winkler elastic foundation with general boundary conditions using Rayleigh–Ritz method. International Journal of Applied Mechanics. 2014;**6**(04):1450043. DOI: 10.1142/S1758825114500434

[11] Damanpack AR, Bodaghi MAHDI, Ghassemi HASSAN, Sayehbani MESBAH. Boundary element method applied to the bending analysis of thin functionally graded plates. Latin American Journal of Solids and Structures. 2013;**10**(3): 549-570. DOI: 10.1590/S1679-78252013000300006

[12] Raju KK, Hinton E. Natural frequencies and modes of rhombic Mindlin plates. Earthquake Engineering and Structural Dynamics. 1980;**8**(1):55-62. DOI: 10.1002/eqe.4290080106

[13] Mindlin RD. Influence of rotary inertia and shear on flexural motions of isotropic elastic plates. The Journal of Applied Mechanics. 1951;**18**(1):31-38. DOI: 10.1115/1.4010217

[14] Liew KM, Xiang Y, Kitipornchai S. Transverse vibration of thick rectangular plates—I. Comprehensive sets of boundary conditions. Computers and Structures. 1993;**49**(1):1-29. DOI: 10.1016/0045-7949(93)90122-T

[15] Della Croce L, Venini P. Finite elements for functionally graded

Reissner–Mindlin plates. Computer Methods in Applied Mechanics and Engineering. 2004;**193**(9–11):705-725. DOI: 10.1016/j.cma.2003.09.014

[16] Chen CS. Non-linear vibration of a shear deformable functionally graded plate. Composite Structures. 2005;**68**(3): 295-302. DOI: 10.1016/j.compstruct.2004.03.022

[17] Fallah A, Aghdam MM, Kargarnovin MH. Free vibration analysis of moderately thick functionally graded plates on elastic foundation using the extended Kantorovich method. Archive of Applied Mechanics. 2013;**83**(2): 177-191. DOI: 10.1007/s00419-012-0645-1

[18] Ganapathi M, Prakash T, Sundararajan N. Influence of functionally graded material on buckling of skew plates under mechanical loads. Journal of Engineering Mechanics. 2006; **132**(8):902-905. DOI: 10.1061/(ASCE)0733-9399(2006)132:8(902)

[19] Shen HS. Non-linear bending response of functionally graded plates subjected to transverse loads and in thermal environments. International Journal of Mechanical Sciences. 2002; **44**(3):561-584. DOI: 10.1016/S0020-7403(01)00103-5

[20] Yang J, Shen HS. Vibration characteristics and transient response of shear-deformable functionally graded plates in thermal environments. Journal of Sound and Vibration. 2002; **255**(3):579-602. DOI: 10.1006/jsvi.2001.4161

[21] Kim YW. Temperature dependent vibration analysis of functionally graded rectangular plates. Journal of Sound and Vibration. 2005;**284**(3–5): 531-549. DOI: 10.1016/j.jsv.2004.06.043

[22] Baferani AH, Saidi AR, Ehteshami H. Accurate solution for free vibration analysis of functionally graded thick rectangular plates resting on elastic foundation. Composite Structures. 2011; **93**(7):1842-1853. DOI: 10.1016/j.compstruct.2011.01.020

[23] Tran LV, Ferreira AJM, Nguyen-Xuan H. Isogeometric analysis of functionally graded plates using higher-order shear deformation theory. Composites Part B: Engineering. 2013;**51**: 368-383. DOI: 10.1016/j.compositesb.2013.02.045

[24] Do TV, Bui TQ, Yu TT, Pham DT, Nguyen CT. Role of material combination and new results of mechanical behavior for FG sandwich plates in thermal environment. Journal of Computational Science. 2017;**21**: 164-181. DOI: 10.1016/j.jocs.2017.06.015

[25] Wang Y, Wu D. Free vibration of functionally graded porous cylindrical shell using a sinusoidal shear deformation theory. Aerospace Science and Technology. 2017;**66**:83-91. DOI: 10.1016/j.ast.2017.03.003

[26] Karama M, Harb BA, Mistou S, Caperaa S. Bending, buckling and free vibration of laminated composite with a transverse shear stress continuity model. Composites: Part B. 1998;**29**:223-234. DOI: 10.1016/S1359-8368(97)00024-3

[27] Attia A, Tounsi A, Adda Bedia EA, Mahmoud S. Free vibration analysis of functionally graded plates with temperature-dependent properties using various four variable refined plate theories. Steel & Composite Structures. 2015;**18**:187-212. DOI: 10.12989/scs.2015.18.1.187

[28] Daouadji TH, Hadji L. Analytical solution of nonlinear cylindrical bending for functionally graded plates.

Geomechanics and Engineering. 2015; **9**(5):631-644. DOI: 10.12989/ GAE.2015.9.5.631

[29] Shahsavari D, Janghorban M. Bending and shearing responses for dynamic analysis of single-layer graphene sheets under moving load. Journal of the Brazilian Society of Mechanical Sciences and Engineering. 2017;**39**(10):3849-3861. DOI: 10.1007/ s40430-017-0863-0

[30] Rachedi MA, Benyoucef S, Bouhadra A, Bachir Bouiadjra R, Sekkal M, Benachour A. Impact of the homogenization models on the thermoelastic response of FG plates on variable elastic foundation. Geomechanics and Engineering. 2020; **22**(1):65-80. DOI: 10.12989/ gae.2020.22.1.065

[31] Ebrahimi F, Barati MR. Hygrothermal effects on vibration characteristics of viscoelastic FG nanobeams based on nonlocal strain gradient theory. Composite Structures. 2017;**159**:433-444. DOI: 10.1016/j. compstruct.2016.09.092

[32] Ghumare SM, Sayyad AS. Analytical solution using fifth order shear and normal deformation theory for FG plates resting on elastic foundation subjected to hygro-thermo-mechanical loading. Materials Today: Proceedings. 2020;**21**: 1089-1093. DOI: 10.1016/j. matpr.2020.01.010

[33] Nebab M, Ait Atmane H, Bennai R, Tahar B. Effect of non-linear elastic foundations on dynamic behavior of FG plates using four-unknown plate theory. Earthquake Structural. 2019;**17**(5): 447-462. DOI: 10.12989/ eas.2019.17.5.447

[34] Akbaş ŞD, Fageehi YA, Assie AE, Eltaher MA. Dynamic analysis of viscoelastic functionally graded porous thick beams under pulse load. Engineering with Computers. 2020;**1-13**: 2. DOI: 10.1007/s00366-020-01070-3

[35] Nguyen HN, Hong TT, Vinh PV, Quang ND, Thom DV. A refined simple first-order shear deformation theory for static bending and free vibration analysis of advanced composite plates. Materials. 2019;**12**(15):2385. DOI: 10.3390/ ma12152385

[36] Shahrjerdi A, Mustapha F, Bayat M, Majid DLA. Free vibration analysis of solar functionally graded plates with temperature-dependent material properties using second order shear deformation theory. Journal of Mechanical Science and Technology. 2011;**25**(9):2195-2209. DOI: 10.1007/ s12206-011-0610-x

[37] Li Q, Iu V, Kou K. Three-dimensional vibration analysis of functionally graded material plates in thermal environment. Journal of Sound and Vibration. 2009;**324**:733-750. DOI: 10.1016/j.jsv.2009.02.036

[38] Reddy JN. Mechanics of Laminated Composite Plates and Shells. Boca Raton, Florida: CRC Press; 2004

[39] Singh VK, Panda SK. Nonlinear free vibration analysis of single/doubly curved composite shallow shell panels. Thin-Walled Structures. 2014;**85**: 341-349. DOI: 10.1016/j.tws.2014.09.003

[40] Selmi A. Exact solution for nonlinear vibration of clamped-clamped functionally graded buckled beam. Smart Structures and Systems. 2020; **26**(3):361-371. DOI: 10.12989/ SSS.2020.26.3.361

[41] Yahea HT, Majeed WI. Free vibration of laminated composite plates in thermal environment using a simple

four variable plate theory. Composite Materials End Engineering. 2021;**3**(3): 179-199. DOI: 10.12989/cme.2021.3.3.179

[42] Attia MA. On the mechanics of functionally graded nanobeams with the account of surface elasticity. International Journal of Engineering Science. 2017;**115**:73-101. DOI: 10.1016/j.ijengsci.2017.03.011

[43] Huang X, Shen H. Non-linear vibration and dynamic response of functionally graded plates in thermal environments. International Journal of Solids and Structures. 2004;**41**(9–10): 2403-2427. DOI: 10.1016/j.ijsolstr.2003.11.012

Chapter 6

# Noise Exposure Assessment in Occupational Settings: Metrics and Correlations

*Hany A. Shawky*

## Abstract

This chapter explores the interplay between various noise exposure metrics and the occupational risks they pose to workers, particularly electrical production workers. Using metrics such as $L_{eq}$, $L_{avg}$, $L_{max}$, and derived ratios ($L_{eq}/L_{avg}$ and $L_{max}/L_{eq}$), the analysis highlights their application in characterizing exposure variability and peak noise impacts. Measurements spanning three industrial sites—metalworking, epoxy processing, and cutting workshops—demonstrate the strengths and limitations of these metrics in quantifying risk and informing hearing conservation efforts.

**Keywords:** noise exposure assessment, occupational noise exposure, exposure metrics, electrical workers, hearing loss

## 1. Introduction

Occupational noise exposure remains a leading cause of Noise-Induced Hearing Loss (NIHL) globally, affecting millions of workers annually [1–5]. In the United States alone, over 22 million workers are exposed to hazardous noise levels, with similar concerns reported in Europe and other industrialized regions. NIHL develops insidiously due to prolonged exposure to excessive noise, causing irreversible damage to the cochlear hair cells [6–12].

This chapter evaluates various metrics for assessing occupational noise exposure. It builds on the premise that traditional metrics like $L_{eq}$ and $L_{avg}$ provide valuable insights into steady-state noise but often fail to adequately address peak noise impacts and variability. Using data collected from three distinct industrial sites, this chapter investigates correlations between standard noise metrics and their derived ratios, offering new perspectives on noise risk assessment.

## 2. Materials and methods

### 2.1 Study sites and participant overview

The study was conducted across three distinct sites in an electrical production factory:

1. Metalworking site: Featuring hydraulic and centric cutting machines.

2. Epoxy site: Divided into chemical treatment and assembling areas.

3. Cutting workshop: Equipped with lathe, forging, and weight-lifting machines.

A total of 99 workers participated in the study. The distribution of workers per site was as follows:

- Metalworking site: 35 workers
- Epoxy site: 30 workers
- Cutting workshop: 34 workers

These workers were monitored over six consecutive days, with exposure intervals of 60, 180, 300, and 480 minutes. Noise exposure was measured using a Type 1 Precision Integrating Sound Level Meter (Bruel & Kjaer Model 2260) calibrated for accuracy.

### 2.2 Noise metrics

Key metrics evaluated included:

- $L_{eq}$: Equivalent continuous sound level over a specified period.
- $L_{avg}$: Average sound level with a 5 dB exchange rate.
- $L_{max}$ and $L_{min}$: Maximum and minimum noise levels recorded within a sampling interval.
- $L_{eq}/L_{avg}$ and $L_{max}/L_{eq}$: Derived ratios quantifying exposure variability and peakiness.

Each metric was measured at a height approximating workers' ear levels, ensuring consistency with international standards (ISO 9612) [13].

### 2.3 Analytical method

Mathematical formulations for noise metric correlations were employed as follow [14]:

Several exposure metrics were logged each minute during every monitored work shift: equivalent sound pressure level during certain period of time $L_{eq}$, using a 3 dB exchange rate, slow response time (1 s) and no threshold; average sound pressure level during certain period of time $L_{avg}$, using a 5 dB exchange rate, slow response time and a 80 dB threshold; and maximum and minimum sound pressure level $L_{max}$, $L_{min}$ with a fast (0.125 s) response time. All levels described in this paper were measured using A-weighting. The $L_{eq}$ and $L_{avg}$ represent average levels integrated over a 1 min period and $L_{max}$ represents the highest maximum level measured during that same period and minimum sound pressure level $L_{min}$. The three metrics defined above ($L_{eq}$, $L_{avg}$ and $L_{max}$) were summarized for a 60, 180, 300 and 480 min (8 h) full shift, during the duration in which a specific tasks were measured, by using the following:

$$L_{ij} = q \log \frac{1}{\sum_k w_{ijk}} \sum_{k=1}^{nij} w_{ijk} 10^{L_{ijk/q}}, \qquad (1)$$

Where $L_{ij}$ is the level for individual i over shift (or task) j summarized over the k = 1 to nij time intervals. The n time intervals were nominally 60, 180, 300 and 480 min (8 h) but where more than one task was reported within a minute, the time-periods were fractions of the minute represented by w. The $L_{eq}$ and $L_{max}$ were calculated using q = 10, while $L_{avg}$ was calculated using q = 16.6 [14]. Note that $L_{max}$ across a sampling period represents the average of the $L_{max}$ levels measured within that period, and as with $L_{eq}$ and $L_{avg}$ levels, averaging was conducted on the exponential scale to appropriately weight the higher-level maxima. These three metrics represent different, widely accepted measures of average levels across the work period Two novel metrics were also developed to represent the variability of exposure. It is well known that $L_{avg}$ and $L_{eq}$ are equivalent in steady state noise, and that their difference becomes larger with increasingly variable noise levels within an Exposure interval. Thus, the ratio of $L_{eq}$ to $L_{avg}$ represents the degree of variability, or intermittency independent of the average exposure level during the time interval. $L_{eq}/L_{avg}$ was thus calculated:

$$\frac{L_{eq}}{L_{avg}} = \frac{1}{\sum_k w_{ijk}} \sum_{k=1}^{nij} w_{ijk} \frac{10^{L_{eq,ijk/10}}}{10^{L_{eq,ijk/10}}}, \qquad (2)$$

Similarly, the degree of 'peakiness' of an exposure can be represented by the ratio of the $L_{max}$ to the $L_{eq}$. The ratio of $L_{max}/L_{eq}$ can be thought of as similar to an average crest factor (ratio of peak to average level) over the exposure interval and was calculated with Eq. (2), exchanging $L_{max}$ for $L_{eq}$ and $L_{eq}$ for $L_{avg}$ In summary, five metrics were compared. $L_{eq}$, $L_{avg}$ and $L_{max}$ indicate average levels of exposure over 60, 180, 300 min. Into full-shift 480 min sampling periods. Two ratio metrics were also developed: $L_{eq}/L_{avg}$ expresses the average variability in levels and increases in increasingly intermittent exposures and $L_{max}/L_{eq}$ expresses the degree of 'peakiness' across the exposure period.

Each of these metrics was described by trade and compared using scatter plots and Pearson's correlation coefficients on an each time scale and for full shifts.

## 3. Results and discussion

### 3.1 Metal working site

Workers at the metal site experienced the highest variability in noise levels. $L_{eq}$ averaged 79.33 dBA over 480 minutes, while $L_{avg}$ was significantly lower at 56.07 dBA. The derived ratio $L_{eq}/L_{avg}$ peaked at 1.58 for shorter intervals, reflecting intermittent exposure. $L_{max}/L_{eq}$ showed minimal variation across all intervals, indicating a consistent peak noise profile (**Figure 1**) [15].

### 3.2 Epoxy site

Noise levels at the epoxy site were influenced by chemical processing activities. $L_{eq}$ values ranged from 67.97 dBA (480 minutes) to 77.01 dBA (60 minutes). The ratio $L_{eq}/L_{avg}$ increased with longer intervals, peaking at 1.52, while $L_{max}/L_{eq}$ remained constant around 1.07.

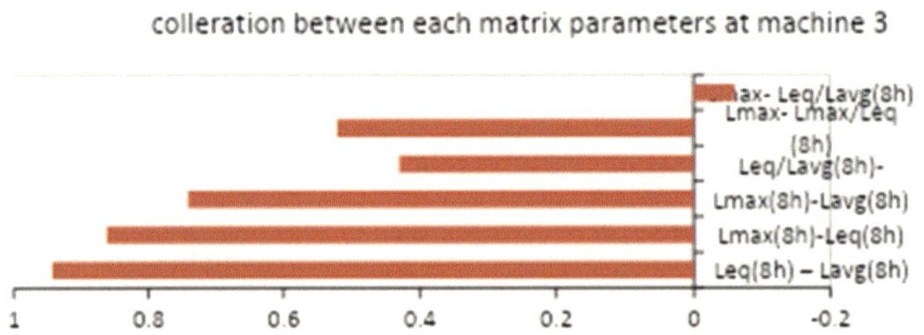


**Figure 1.**
*Exposure variability ($L_{eq}/L_{avg}$) across different time intervals in the cutting machines in the workshop.*

## 3.3 Cutting workshop

Cutting machines and forging activities generated the highest peak noise levels, with $L_{max}$ reaching 92.12 dBA. Unlike the other sites, $L_{max}/L_{eq}$ values showed slight variation, reflecting distinct exposure profiles for tasks involving impact-type noise.

**Figure 2** illustrates the strong correlation (r = 0.96) between $L_{eq}$ and $L_{avg}$ for workers in the metal site over an 8-hour shift. Despite the correlation, significant differences in absolute values highlight variability due to the differing exchange rates (3 vs. 5 dB).

The ratio of $L_{eq}/L_{avg}$ demonstrates increased exposure variability with longer intervals (60, 180, 300, and 480 minutes). This trend underscores the intermittent nature of chemical treatment activities **(Figure 3)**.

**Figure 1** presents the “peakiness” of noise exposure, represented by the ratio $L_{max}/L_{eq}$, for cutting machines in the workshop. Despite the high noise levels recorded, the $L_{max}/L_{eq}$ ratio remains relatively constant, indicating consistent peak characteristics during different tasks.

**Figure 1** shows minimal variability in $L_{max}/L_{eq}$ values across sampling intervals, indicating consistent peak noise levels for cutting, forging, and lathe machines. Peaks remained steady at approximately 1.01, even during short-term measurements.

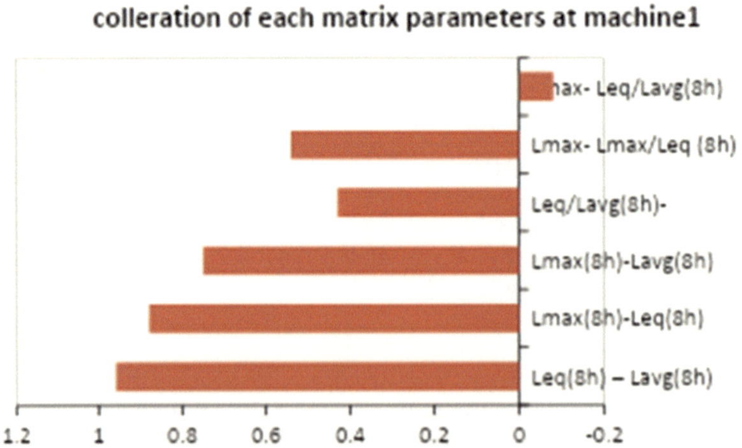


**Figure 2.**
*Correlation between $L_{eq}$ and $L_{avg}$ over an 8-hour exposure period in the metal site.*

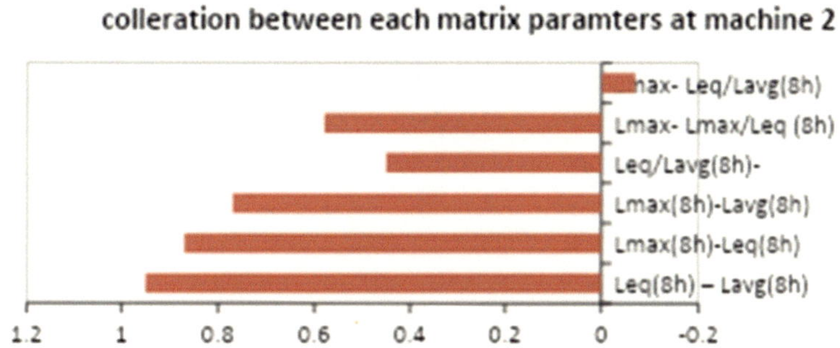


**Figure 3.**
*Exposure variability ($L_{eq}/L_{avg}$) across different time intervals in the epoxy site.*

**Table 1** summarizes the noise metrics for workers in the metal site across intervals of 60, 180, 300, and 480 minutes. Variability in exposure levels is captured by the $L_{eq}/L_{avg}$ ratio, which is highest during shorter intervals.

**Table 2** highlights the noise levels at two key positions in the epoxy site. The chemical treatment area shows greater variability and higher $L_{eq}$ values compared to the assembling area, reflecting more fluctuating and intense noise levels.

**Table 3** presents noise exposure data for key machines in the cutting workshop. Among all equipment, the lathe machine exhibited the highest peak noise levels ($L_{max}$).

**Table 4** provides an overview of the derived variability metrics ($L_{eq}/L_{avg}$ and $L_{max}/L_{eq}$) for all sites and intervals. The highest variability was observed in the epoxy site's chemical treatment area during the longest interval (480 minutes).

**Table 4** presents $L_{eq}$, $L_{avg}$, $L_{max}$, and derived ratios ($L_{eq}/L_{avg}$ and $L_{max}/L_{eq}$) for the metalworking site. Notably, exposure variability ($L_{eq}/L_{avg}$) was highest during the shortest time interval (60 minutes), reflecting intermittent noise patterns.

| n (min) | $L_{eq}$ (SD) | $L_{avg}$ (SD) | $L_{max}$(SD) | $L_{min}$ (SD) | $L_{eq}/L_{avg}$ (SD) | $L_{max}/L_{eq}$ (SD) |
|---|---|---|---|---|---|---|
| 60 | 88.36 (0.53) | 56.07 (0.7) | 98.34 (1.35) | 81.98 (1.59) | 1.58 (1.24) | 1.11(2.12) |
| 180 | 83.59 (0.68) | 63.14 (3.35) | 93.57 (1.32) | 77.21(1.46) | 1.32 (1.54) | 1.12 (2.41) |
| 300 | 81.38 (0.72) | 59.46 (3.59) | 91.35 (1.34) | 74.99 (1.46) | 1.37 (2.31) | 1.12 (1.39) |
| 480 | 79.33 (0.72) | 56.07 (4.05) | 89.31 (1.22) | 72.95 (1.33) | 1.41(1.11) | 1.13 (2.12) |

**Table 1.**
*Noise metrics across different time intervals in the metal site.*

| n (min) | $L_{eq}$ (SD) | $L_{avg}$ (SD) | $L_{max}$(SD) | $L_{min}$ (SD) | $L_{eq}/L_{avg}$ (SD) | $L_{max}/L_{eq}$ (SD) |
|---|---|---|---|---|---|---|
| 60 | 92.42 (0.94) | 75.09 (2.69) | 106.38 (5.68) | 98.63 (8.12) | 1.23 (3.38) | 1.15 (3.04) |
| 180 | 87.65 (0.89) | 67.17 (3.14) | 101.61(6.23) | 93.86 (8.12) | 1.30 (5.53) | 1.16 (3.59) |
| 300 | 85.43 (0.83) | 63.49 (3.86) | 99.4 (5.23) | 91.64 (8.54) | 1.35 (5.31) | 1.16 (3.11) |
| 480 | 83.39 (3.09) | 60.10 (4.09) | 97.35 (5.7) | 89.6 (9.97) | 1.39 (4.24) | 1.17 (4.44) |

**Table 2.**
*Noise metrics at two positions in the epoxy site.*

| n (min) | $L_{eq}$ (SD) | $L_{avg}$ (SD) | $L_{max}$(SD) | $L_{min}$ (SD) | $L_{eq}/L_{avg}$ (SD) | $L_{max}/L_{eq}$ (SD) |
|---|---|---|---|---|---|---|
| 60 | 87.32 (1.58) | 70.02 (1.04) | 95.90 (1.99) | 71.60 (0.88) | 1.25 (2.68) | 1.10 (0.1) |
| 180 | 82.54 (1.47) | 62.10 (2.21) | 91.13 (1.99) | 75.86 (0.77) | 1.33 (4.65) | 1.10 (1.0) |
| 300 | 80.33 (1.39) | 58.41(5.09) | 88.91(1.88) | 73.64 (0.92) | 1.38 (3.17) | 1.11 (1.56) |
| 480 | 78.28 (2.21) | 55.02 (2.04) | 86.87 (1.79) | 71.6 (0.87) | 1.42 (3.74) | 1.11 (2.83) |

**Table 3.**
*Noise metrics for machines in the cutting workshop.*

| Site | Intervals (minutes) | $L_{eq}/L_{avg}$ (mean ± SD) | $L_{max}/L_{eq}$ (mean ± SD) |
|---|---|---|---|
| Metal site | 480 | 1.41 ± 1.11 | 1.13 ± 2.12 |
| Epoxy site 1 | 480 | 1.52 ± 2.02 | 1.07 ± 5.90 |
| Cutting Workshop | 480 | 1.43 ± 0.97 | 1.01 ± 1.11 |

**Table 4.**
*Variability metrics across time intervals for all sites.*

| Machine | Interval (minutes) | $L_{eq}$ (mean ± SD, dBA) | $L_{avg}$ (mean ± SD, dBA) | $L_{max}$ (mean ± SD, dBA) | $L_{eq}/L_{avg}$ (mean ± SD) | $L_{max}/L_{eq}$ (mean ± SD) |
|---|---|---|---|---|---|---|
| Lathe machine | 60 | 91.42 ± 0.20 | 74.10 ± 0.33 | 92.12 ± 0.08 | 1.23 ± 0.07 | 1.01 ± 0.13 |
| Forging machine | 480 | 82.69 ± 0.15 | 59.41 ± 0.66 | 83.26 ± 0.15 | 1.39 ± 1.06 | 1.01 ± 0.59 |
| Cutting machine | 300 | 84.65 ± 0.25 | 62.72 ± 3.17 | 85.18 ± 0.25 | 1.35 ± 1.67 | 1.01 ± 0.49 |
| Weight-lifting machine | 480 | 77.14 ± 0.61 | 53.88 ± 0.63 | 78.02 ± 0.68 | 1.43 ± 0.97 | 1.01 ± 1.11 |
| Forging machine | 480 | 82.69 ± 0.15 | 59.41 ± 0.66 | 83.26 ± 0.15 | 1.39 ± 1.06 | 1.01 ± 0.59 |
| Cutting machine | 300 | 84.65 ± 0.25 | 62.72 ± 3.17 | 85.18 ± 0.25 | 1.35 ± 1.67 | 1.01 ± 0.49 |

**Table 5.**
*Summary of noise exposure metrics ($L_{eq}$, $L_{avg}$, $L_{max}$) and derived ratios ($L_{eq}/L_{avg}$ and $L_{max}/L_{eq}$) for different machines in the cutting workshop.*

**Table 5** summarizes noise exposure data for different machines in the cutting workshop. Among all machines, the lathe machine exhibited the highest peak noise levels.

**Figure 4** illustrates the weak correlation between peak noise levels ($L_{max}$) and the exposure variability ratio ($L_{eq}/L_{avg}$) in the metal site. The results suggest that peak noise levels do not significantly affect the overall variability metric.

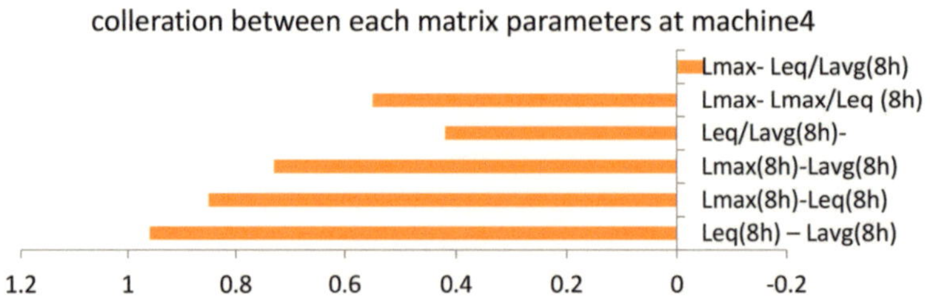


**Figure 4.**
*Relationship between $L_{max}$ and $L_{eq}/L_{avg}$ in metal site.*

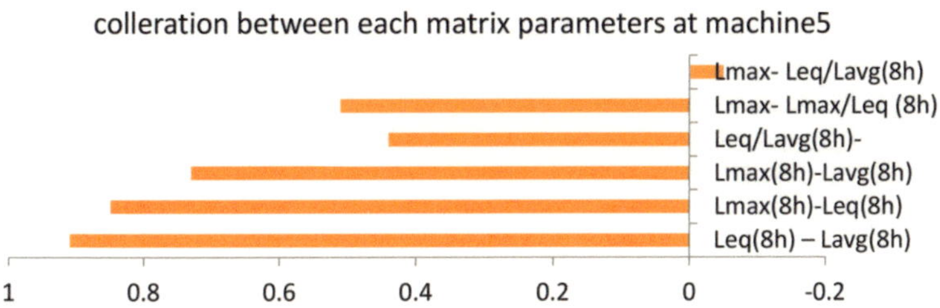


**Figure 5.**
*Comparative noise levels in the epoxy and cutting sites.*

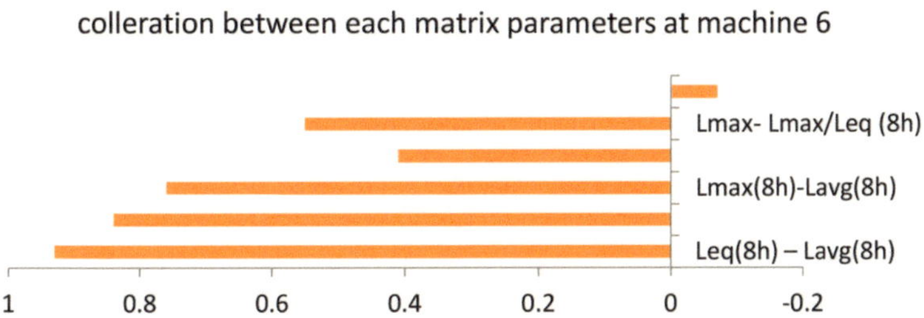


**Figure 6.**
*Comparative noise levels across sites at machine 6.*

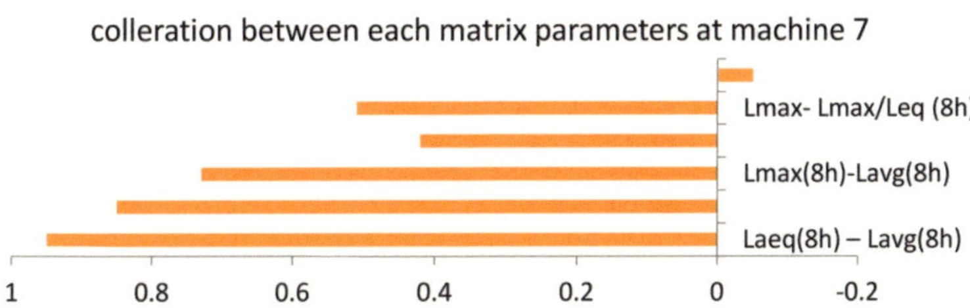


**Figure 7.**
*Comparison of $L_{max}$ across sites at machine 7.*

**Figure 5** compares noise levels across different time intervals in the epoxy and cutting sites. The cutting site demonstrated consistently higher $L_{max}$ values, reflecting the impact of mechanical tasks like forging and lathe operations.

**Figures 6** and 7 compare maximum noise levels ($L_{max}$) across all three sites for an 8-hour exposure period. The cutting site consistently recorded the highest peaks due to impact noise activities, such as forging and cutting.

## 4. Conclusion

This chapter highlights the strengths and limitations of standard noise exposure metrics in occupational settings. By incorporating derived ratios, such as $L_{eq}/L_{avg}$, it is possible to better characterize exposure variability and peak noise effects. These findings have significant implications for refining noise assessment methodologies and improving hearing conservation programs.

## Conflict of interest

The author declares no conflict of interest.

## Author details

Hany A. Shawky
Acoustic Department, Mass and Force Metrology Division, National Institute of Standard (NIS), Egypt

*Address all correspondence to: hany.amir15@gmail.com

## References

[1] Arezes PM, Miguel AS. Hearing protection use in industry: The role of risk perception. Safety Science. 2005;**43**(4):253-267

[2] CDC. Noise and Hearing Loss Prevention. 2015. Available from: http://www.cdc.gov/niosh/topics/noise/stats.html [Accessed: 6 January 2016]

[3] Dunn DE. Noise. In: Levy BS, Wegman DH, editors. Occupational Health: Recognizing and Preventing Work Related Disease and Injury. 4th ed. Philadelphia: Lippincott Williams & Wilkins; 2000. pp. 367-377

[4] May J. Occupational hearing loss. American Journal of Industrial Medicine. 2000;**37**:112-120

[5] Aybek A, Kamer HA, Arslan S. Personal noise exposures of operators of agricultural tractors. Applied Ergonomics. 2010;**41**(2):274-281

[6] Fernandez MD, Quintana S, ChavarriaN, BallesterosJA. Noiseexposure of works of the construction sector. Applied Acoustics. 2009;**70**:753-760

[7] Masterson EA, Themann CL, Luckhaupt SE, Li J, Calvert GM. Hearing difficulty and tinnitus among U.S. workers and non-workers in 2007. American Journal of Industrial Medicine. Apr 2016;**95**(4)

[8] Ljungberg JK, Neely G. Stress, subjective experience and cognitive performance during exposure to noise and vibration. Journal of Environmental Psychology. 2007;**27**(1):44-54

[9] Atherley GR, Martin AM. Equivalent-continuous noise level as a measure of injury from impact and impulse noise. The Annals of Occupational Hygiene. 1971;**14**:11-23

[10] Dancer A, Buck K, Parmentier G, et al. The specific problems of noise in military life. Scandinavian Audiology. Supplementum. 1998;**48**:123-130

[11] OSHA. Occupational Noise Exposure: Hearing Conservation Amendment; Final Rule. Fed Reg. Vol. 48. Washington, DC: US Department of Labor, Occupational Safety and Health Administration; 1983. pp. 738-785

[12] NIOSH. Criteria for a Recommended Standard: Occupational Noise Exposure. Cincinnati, OH: US Department of Health and Human Services/Centers for Disease Control, National Institute for Occupational Safety and Health; 1998

[13] ISO 9612. Acoustics- Determination of Occupational Noise Exposure Engineering Method. 2nd ed. Switzerland: International Standard; 2009

[14] Earshen J. Sound measurement: Instrumentation and noise descriptors. In: Berger E, Royster L, Royster J, Driscoll D, Layne M, editors. The Noise Manual. Fairfax, VA: American Industrial Hygiene Association; 2000

[15] Shawky HA. Study relationship between noise metrics parameters exposure among electrical production workers. Romanian Journal of Acoustics and Vibration. 2021;**18**(2). ISSN 1584-7284